The Introduction to Private Cloud using Oracle Exadata and Oracle Database

The Introduction to Private Cloud using Oracle Exadata and Oracle Database

Okcan Yasin Saygili

CRC Press
Taylor & Francis Group
Boca Raton London New York

CRC Press is an imprint of the
Taylor & Francis Group, an **informa** business

CRC Press
Taylor & Francis Group
6000 Broken Sound Parkway NW, Suite 300
Boca Raton, FL 33487-2742

First issued in paperback 2020

ISBN 13: 978-0-367-67034-4 (pbk)
ISBN 13: 978-0-367-07462-3 (hbk)

Visit the Taylor & Francis Web site at
http://www.taylorandfrancis.com

and the CRC Press Web site at
http://www.crcpress.com

Contents

Preface

The products described in this book are actually divided into two parts, one of which is Exadata and the other is Oracle's own database. These two products are handled separately; each one has been described independently in the book. For instance, all the database structures in the book were tested and written in Oracle Database (software only) and tested in virtual machine, not on the database in Exadata. In examples shown, because of this issue, performance loss in the database or a different expression is not related to Exadata and is directly related to the database in the virtual machine. As I said earlier, I have explained Exadata hardware and database separately. When writing the database section, I didn't describe directly Exadata's database, but rather I wrote about the database software only in a virtual system.

Cloud Computing

1

The invention of the computer was actually the beginning of a new era. It was the precursor to a data repository that had not evolved in the earlier part of this era. When the repository for data was originally invented, no one knew how important its use would be in the future. Storing data, which has been one of the two most important functions since the advent of information technology (IT), is likely to keep us more occupied in the future. The storage and processing of data is a costly affair. The cost of processing and maintenance is significant when the companies store data in their data centers, and mostly companies would prefer to avoid such situations. Cloud computing is a new technology of this era that is growing and expanding day by day. Today, data of media, such as Dropbox and Instagram, and also all the enterprise data are stored in the cloud. Nowadays all companies are investigating how to store their data for future use, but some companies already have these data in place. In this trend, the IT industry seeks to move on-premise solutions to the cloud. The advantage of cloud is the saving in cost achieved by the use of solutions. Therefore, IT projects can be initialized at minimum cost. On the other hand, there is no cost saving with on-premise solutions as they are purchased. However, most of the time, these solutions aren't estimated correctly or used completely. A project may be started without expensive hardware and software and then updated to meet new requirements. Cloud solutions provide an opportunity to a team's members to work together, where each member of the team can, for example, easily connect to a cloud account for editing, copying, and deleting team documents on the cloud service. In general, cloud services are of two types, namely, public and private cloud. If everyone can access a cloud service, it is a public cloud. If the access to the cloud service is restricted to an organization, it is a private cloud. This book discusses about private cloud service.

Cloud is the most important technology in the world of informatics. The IT sector is growing day by day, and hence, the cloud data is also increasing. At this point, where to store the data and how to manage it under a single roof are the problems to be addressed; moreover, the processing of data is a separate and critical problem of the sector.

One of the most important topics of my work is cloud technologies and Gartner's estimated cloud technologies, which predicted that the market value of cloud technologies will increase from $209 billion in 2016 to $246.8 billion in 2020. It is not usually possible to see such a big market (Facebook, Instagram, etc.).

The management and maintenance of heterogeneous systems is very difficult. Therefore, the architecture and data of the system must be created and managed very well. So far, cloud systems primarily have been the backbone of a company, and they will continue to be so in the future as well [1,2].

PROBLEMS IN ON-PREMISE TECHNOLOGIES

Security Problems in On-Premise Technologies

Breach of security, especially information security, is one of the issues that we increasingly see in Turkey. As we are in the initial stages of providing better security, understanding some basic concepts is important. These basic concepts are the subject of this book. The lack of understanding of these basic concepts leads to solutions that provide security starts from the beginning. These solutions are mostly short term and cannot achieve success in the long term. In places where there is no cloud technology, it is more difficult to maintain security [1,2].

Cost Problems in On-Premise Technologies

The most expensive item in the IT sector is the employee, and this is the main reason for the transition to cloud technologies.

Effect of Natural Life in On-Premise Technologies

Cloud technologies in certain places, especially in distributed form of different types of software, work on servers that have been damaged to a certain extent during their operational life. These systems consume more electricity than required, and the cooling of unnecessary servers in operation involves additional costs.

IT Problems in On-Premise Technologies

It is not possible to talk about cloud systems in important instances such as disaster recovery, backup solutions, and 100% working guarantee.

Problems in Switching from On-Premise Technologies to Cloud Technologies

Switching to cloud systems is seriously expensive in terms of both time and cost, and if these transitions are not planned correctly, they could even lead to closure of the project. The transition should be well planned to avoid extra cost, which can be both opex and capex.

ADVANTAGES OF CLOUD

Cloud technology enables us to manage all solutions on remote servers. For those planning to adopt cloud technologies, there are many advantages to consider, including the following:

- *Cost*: Cloud vendors offer services at a reasonable price, and customers can easily pay through cloud web sites. For example, Oracle Cloud can be bought through online web sites.
- *Flexibility*: Users can access cloud technologies from wherever they are, and this is the most popular reason for using cloud services.
- *Customer support*: Cloud services provide high-quality support or solutions to customers' problems.
- *Maintenance*: Cloud provides a specific service-level agreement (SLA) to customers, particularly for hardware replacements.

CLOUD ARCHITECTURAL MODELS

Cloud vendors provide three cloud services: private, hybrid, and public.

Private cloud: This is the cloud technology in which only users within an organization can have access and connect to it from outside through links.

Public cloud: This is a cloud architecture that can be accessed by both outside and inside users.

Hybrid cloud: It is a cloud architecture that is open to users both internally and externally.

CLOUD SERVICES

Cloud services vary according to their applications and the services they provide, so they are named differently.

SaaS (Software as a Service): This is a subscription-based model for cloud application. This model provides applications support, and it can provide support for sales, marketing, and human resources in companies.

PaaS (Platform as a Service): This is a subscription-based model for developing new applications. It is a new-generation cloud that is not just an infrastructure but also provides new development technologies. This model provides rapid development in service through PaaS technologies.

IaaS (Infrastructure as a Service): This is a subscription-based model for using hardware. IaaS provides new-generation hardware for developing new services.

SUMMARY

Data centers are being managed in many companies, including some of the Fortune 500, and form the backbone of critical applications and business operations, so cloud computing has always been a high priority for the data center, as keeping a business going 24/7, 365 days a year, is vital. This chapter discusses the options for cloud and evolution of cloud in time.

Relational Database Management System (RDBMS)

2

Database systems have been used in the history of computers to store large amount of data in organizations or to reduce the data loss in the program structures.

Today, database systems are available on the market. This is because creating systems is no longer important. A database system is just a software that stores data. There are a lot of databases, but what is missing is a program that can manage and modify data efficiently and quickly. A database system is at the top of the data systems, achieving good speed over its use. Providing access to data when needed is not enough; achieving the desired speed is equally important. Considering that there are so many features in databases, the answer is simple: to increase only the speed of the system and better understand the developer. A mess in a database system is an unbearable pain for many developers, so it is so important that the system is clean. The software sector is very large, and there are certain branches that feed this sector. As the software sector grows, but only the development and interfaces zones are not grow alone and also other IT zones grow with it. Because a software application produces data in a certain way, the data needs to be stored in a certain way. In fact, these things did not appear to be very important at first, but after few years, the amount of the data increased. The increase was tremendous and made storing the data normally in a simple file impossible and data transfer between the files very difficult. Many disagreements on the storage began to occur. Some database software applications were developed, but one of the solutions closest to the structure we are currently using is IBM.

HISTORY OF RDBMS

Relational database is now seen as the core of most software, and this system has been evolving and in use since 1970. Currently, billions of software and devices are working on this application. One of the founders of this application is E. F. (Tedd) Codd. In 1974, IBM first developed a data storage project, System R. In 1979, Oracle introduced its first commercial database project. After this project, many databases were developed; among them were Sybase ASE, DB2, Informix, PostgreSQL, and MySQL. This developed software was storing not only the data but also the workload of the applications on the software. Thanks to this software, many large-scale companies, hospitals, and factories can their data.

The phrase "relational database" was introduced in IBM by E. F. Codd in 1970. Codd wrote an article titled "A Relational Model of Data for Large Shared Data Banks" and aroused great repercussions. But there were 12 rules issued by Codd that made RDBMS more understandable.

The data was retained in the so-called table for users, which had a series of rows and columns. Major universities using this model were as follows:

1. **Massachusetts Institute of Technology (1971)**
2. **Michigan University (1969)**

The first system sold was Multics Relational Data Store with RDBMS. Following this, Ingres and IBM BS12 were commercially available. From 2008, many commercial RDBMSs have been using SQL as the query language.

RDBMS

E. F. Codd described briefly about RDBMS: "Future users of large data banks must be protected from having to know how the data is organized in the machine (the internal representation). A prompting service which supplies such information is not a satisfactory solution. Activities of users at terminals and most application programs should remain unaffected when the internal representation of data is changed and even when some aspects of the external representation are changed. Changes in data representation will often be needed as a result of changes in query, update, and report traffic and natural

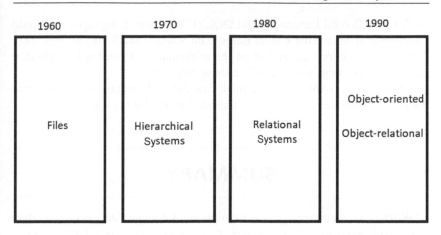

growth in the types of stored information. Data has been kept in many formats for the user." [3]. It is an example of a relational database structure in which data are appropriately stored in tables and interact when they are processed later [7–11].

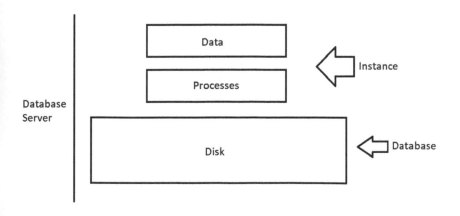

SQL

SQL stands for structured query language, and its main purpose is to manage and design RDBMSs. This language has a few advantages: It is a DML (Data Manipulation Language), DDL (Data Definition Language),

DCL (Data Control Language), and DQL (Data Query Language). E.F. Codd has described these in the article "Large Data Shared Data Banks for SQL" [11]. Although not entirely dependent on the relational model defined by Codd, it has become the most widely used database language.

SQL ANSI (American National Standards Institute) standards were adopted in 1986 and 1987 and were accepted by the ISO [14–18].

SUMMARY

RDBMS is a data management system provides database access to more than one person and stores a large amount of data of different types. It is a database structure that can process the data in the database enabling them to interact with each other and maintain the accuracy of the structure. This chapter explains the fundamentals of RDBMS, and the later chapters explain more about database features.

Private Cloud Database and Pluggable Database

3

Today, the amount of stored data is growing day by day, and therefore the complexity in their management is also increasing every day; in other words, data's future is changing with this condition. The biggest problem these days is determining where this data goes and how it can be managed. RDBMS was designed to handle this problem. The usage of databases has increased over the past years, and now dozens of databases are used in data centers to store this data. However, the complexity in the management of these systems normally increases as the data increases, and consolidation of these data under one roof is required. Older silo solutions no longer support these data sizes. In fact, a lot of time is spent and resources are used due to separate systems in a data center because many databases are managed at different servers and a database administrator (DBA) and technical person must be assigned to every machine; in other words, a new resource must be provided for every database. Therefore, now companies have a better understanding of this case and try to move all their data to one database and one system for consolidation [19].

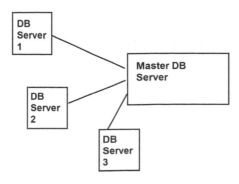

PLUGGABLE DATABASE

Oracle 12c was released as a pluggable database (PDB) to host one or more databases from only one database server through a container database (CDB). Therefore, all systems are managed from only one point with a CDB and PDBs. Generally, this is known as multitenant architecture and includes an Oracle database working as a container database [20]. A CDB includes PDBs. This portable database collects schemas, objects, and nonschematic objects and works with Oracle Net client as a non-CDB. This feature was incorporated in Oracle 12.1.1 version [21].

Every CDB has the following containers:

Root system: The root system includes metadata and specific users. This metadata contains PL/SQL packages with specified source codes. A specific user is a database user in every container. This system is called CDB$ROOT.

One seed PDB: The seed PDB includes the template which can be used by the CDB to create new PDBs.

More PDBs: PDBs are entities created by users, and they contain data and codes required for a set of features [22,23].

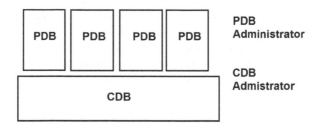

ADVANTAGES OF PDB

1. Cost reduction

Consolidation is a reality in IT sectors, and it has a direct impact on IT costs. It covers both hardware and software and thus helps to avoid unnecessary consumption, especially of hardware. Companies can save hardware and software costs through consolidation.

2. Isolation

CDB and PDBs are totally isolated from each other; in other words, there is no effect on the system while performing any operation on CDB.

3. Patching and Upgrading

Consolidation impacts not only cost but also the workload of technicians. Patching and upgrading are really a problem; every database performs patch script, but this PDB performs a one-time patching and upgrading for all databases.

4. Management

The CDB manages systems by executing a single operation, for example, recovery manager (RMAN) backup, patching, or upgrading [24–27].

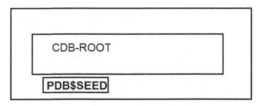

Now take a look at how to use this PDB [28,29].

1. Show Oracle Database Name Condition with below query.
```
SQL> COLUMN NAME FORMAT A8
SQL> SELECT NAME, CON _ ID, DBID, CON _ UID, GUID
   FROM V$CONTAINERS ORDER BY CON _ ID;
NAME CON _ ID DBID CON _ UID GUID
-------- ---------- ---------- ---------- --------
------------------------
CDB$ROOT 1 1425867235 1 FD9AC20F64D344D7E043B6A9
   E80A2F2F
PDB$SEED 2 3130831091 3130831091 2689AFEA555B1A16
   E0535405040A5905
PDBORCL 3 679092436 679092436 2689DFC1DDE01CFAE053
   5405040A4654
```
2. Show database condition list with below query. This query shows the status of PDBs.
```
SQL> COLUMN NAME FORMAT A15
COLUMN RESTRICTED FORMAT A10
COLUMN OPEN _ TIME FORMAT A30
SELECT  NAME,  OPEN _ MODE,  RESTRICTED,  OPEN _
   TIME FROM V$PDBS;SQL> SQL> SQL> SQL>
```

```
NAME OPEN _ MODE RESTRICTED OPEN _ TIME
--------------- ---------- ---------- ------------
-------------------
PDB$SEED  READ  ONLY  NO  10-DEC-15  12.50.37.124
  PM +02:
00
PDBORCL  READ  WRITE  NO  10-DEC-15  12.45.51.424
  PM +02:
00
```

3. Show table name with below query.

```
SQL> COLUMN PDB _ NAME FORMAT A15
COLUMN OWNER FORMAT A15
COLUMN TABLE _ NAME FORMAT A30
SQL>
SQL> SELECT p.PDB _ ID, p.PDB _ NAME, t.OWNER,
  t.TABLE _ NAME
FROM dba _ PDBS p, CDB _ TABLES t
WHERE p.PDB _ ID > 2 AND
t.OWNER IN('HR','OE') AND
p.PDB _ ID = t.CON _ ID
ORDER BY p.PDB _ ID; 2 3 4 5 6
PDB _ ID PDB _ NAME OWNER TABLE _ NAME
---------- --------------- --------------- --------
----------------------
3 PDBORCL HR JOB _ HISTORY
3 PDBORCL HR COUNTRIES
3 PDBORCL HR JOBS
3 PDBORCL HR DEPARTMENTS
3 PDBORCL HR LOCATIONS
3 PDBORCL HR REGIONS
3 PDBORCL OE SUBCATEGORY _ REF _ LIST _ NESTEDTAB
3 PDBORCL OE PRODUCT _ REF _ LIST _ NESTEDTAB
3 PDBORCL OE PROMOTIONS
3 PDBORCL OE PRODUCT _ DESCRIPTIONS
3 PDBORCL OE PRODUCT _ INFORMATION
PDB _ ID PDB _ NAME OWNER TABLE _ NAME
---------- --------------- --------------- --------
----------------------
3 PDBORCL OE INVENTORIES
3 PDBORCL OE ORDERS
3 PDBORCL OE ORDER _ ITEMS
3 PDBORCL OE WAREHOUSES
```

```
3 PDBORCL OE CUSTOMERS
3 PDBORCL HR EMPLOYEES
17 rows selected.
```
4. Show users in database with below query.
```
SQL> COLUMN PDB _ NAME FORMAT A15
COLUMN USERNAME FORMAT A30
SELECT p.PDB _ ID, p.PDB _ NAME, u.USERNAME
FROM dba _ PDBS p, CDB _ USERS u
WHERE p.PDB _ ID > 2 AND
p.PDB _ ID = u.CON _ ID
ORDER BY p.PDB _ ID;
SQL> SQL> SQL> 2 3 4 5
PDB _ ID PDB _ NAME USERNAME
---------- --------------- -----------------------
   -------
3 PDBORCL BI
3 PDBORCL SYS
3 PDBORCL IX
3 PDBORCL SH
3 PDBORCL OE
3 PDBORCL HR
3 PDBORCL SCOTT
3 PDBORCL ORACLE _ OCM
3 PDBORCL OJVMSYS
3 PDBORCL SYSKM
3 PDBORCL XS$NULL
PDB _ ID PDB _ NAME USERNAME
43 rows selected.
```
5. Show PDB data files with below query.
```
SQL> COLUMN PDB _ ID FORMAT 999
COLUMN PDB _ NAME FORMAT A8
COLUMN FILE _ ID FORMAT 9999
COLUMN TABLESPACE _ NAME FORMAT A10
COLUMN FILE _ NAME FORMAT A45
SELECT   p.PDB _ ID,   p.PDB _ NAME,   d.FILE _ ID,
  d.TABLESPACE _ NAME, d.FILE _ NAME
FROM dba _ PDBS p, CDB _ DATA _ FILES d
WHERE p.PDB _ ID = d.CON _ ID
ORDER BY p.PDB _ ID;SQL> SQL> SQL> SQL> SQL> SQL>
  2 3 4
PDB _ ID PDB _ NAME FILE _ ID TABLESPACE FILE _
  NAME
```

```
------ -------- ------- ---------- ---------------
-------------------------------
3 PDBORCL 8 SYSTEM /u01/app/oracle/oradata/orcl/
  pdborcl/system01
.dbf
3  PDBORCL  11  EXAMPLE  /u01/app/oracle/oradata/
  orcl/pdborcl/example0
1.dbf
3 PDBORCL 10 USERS /u01/app/oracle/oradata/orcl/
  pdborcl/SAMPLE _ S
CHEMA _ users01.dbf
3 PDBORCL 9 SYSAUX /u01/app/oracle/oradata/orcl/
  pdborcl/sysaux01
.dbf
PDB _ ID    PDB _ NAME    FILE _ ID    TABLESPACE
FILE _ NAME
------ -------- ------- ---------- ---------------
-------------------------------
```

6. Show temp file from CDB with below query.

```
SQL> COLUMN CON _ ID FORMAT 999
COLUMN FILE _ ID FORMAT 9999
COLUMN TABLESPACE _ NAME FORMAT A15
COLUMN FILE _ NAME FORMAT A45
SELECT  CON _ ID,  FILE _ ID,  TABLESPACE _ NAME,
  FILE _ NAME
FROM CDB _ TEMP _ FILES
ORDER BY CON _ ID;SQL> SQL> SQL> SQL> SQL> 2 3
CON _ ID FILE _ ID TABLESPACE _ NAME FILE _ NAME
------ ------- --------------- --------------------
-------------------------
1 1 TEMP /u01/app/oracle/oradata/orcl/temp01.dbf
3  3  TEMP  /u01/app/oracle/oradata/orcl/pdborcl/
  pdborcl _
temp012015-12-10 _ 12-43-14-PM.dbf
```

7. Show Oracle service name from CDB with below query.

```
SQL> COLUMN NETWORK _ NAME FORMAT A30
COLUMN PDB FORMAT A15
COLUMN CON _ ID FORMAT 999
SELECT PDB, NETWORK _ NAME, CON _ ID FROM CDB _
  SERVICES
WHERE PDB IS NOT NULL AND
CON _ ID > 2
```

```
ORDER BY PDB;SQL> SQL> SQL> SQL> 2 3 4
PDB NETWORK _ NAME CON _ ID
------------- --------------------------- ------
PDBORCL pdborcl.localdomain 3
```
8. Show container id and container name with below query.
```
SQL> SHOW CON _ ID
CON _ ID
------------------------------
1
SQL> SHOW CON _ NAME
CON _ NAME
------------------------------
CDB$ROOT
```
9. Show database name and PDB relation with below command.
```
SQL> COLUMN DB _ NAME FORMAT A10
COLUMN CON _ ID FORMAT 999
COLUMN PDB _ NAME FORMAT A15
COLUMN OPERATION FORMAT A16
COLUMN OP _ TIMESTAMP FORMAT A10
COLUMN CLONED _ FROM _ PDB _ NAME FORMAT A15
SELECT DB _ NAME, CON _ ID, PDB _ NAME, OPERATION,
  OP _ TIMESTAMP, CLONED _ FROM _ PDB _ NAME
FROM CDB _ PDB _ HISTORY
WHERE CON _ ID > 2
ORDER BY CON _ ID;SQL> SQL> SQL> SQL> SQL> SQL>
  SQL> 2 3 4
DB _ NAME  CON _ ID  PDB _ NAME  OPERATION  OP _
  TIMESTA CLONED _ FROM _ PDB
---------- ------ --------------- ----------------
---------- ---------------
SEEDDATA   3   SAMPLE _ SCHEMA   CREATE   07-JUL-14
  PDB$SEED
ORCL 3 PDBORCL PLUG 10-DEC-15 PDBORCL
SEEDDATA 3 SAMPLE _ SCHEMA UNPLUG 07-JUL-14
```
10. Show Oracle database service condition with below query.
```
[oracle@rac1 ~]$ ps -aef|grep pmon
oracle 7152 1 0 12:39 ? 00:00:01 ora _ pmon _ orcl
oracle 9445 8663 0 15:43 pts/4 00:00:00 grep pmon
Show Oracle database listener path
[oracle@rac1 ~]$ adrci
ADRCI: Release 12.1.0.2.0 - Production on Thu
  Dec 10 15:44:30 2015
```

```
Copyright   (c)   1982,   2014,   Oracle   and/or   its
   affiliates. All rights reserved.
ADR base = "/u01/app/oracle"
adrci> show homes
ADR Homes:
diag/rdbms/orcl/orcl
diag/tnslsnr/rac1/listener
Check PDB with below command
SQL> select con _ id, dbid, name
from v$pdbs; 2
CON _ ID DBID NAME
---------- ---------- ----------------------------
2 3130831091 PDB$SEED
3 679092436 PDBORCL
```

11. Show service name with below query.

```
SQL> show parameter service
NAME TYPE VALUE
------------------------------------   -----------
------------------------------
service _ names string orcl.localdomain
SQL>
```

12. Show database info with below query.

```
SQL> select con _ id, name, open _ time, cre-
   ate _ scn, total _ size from v$pdbs;
CON _ ID NAME
---------- ------------------------------
OPEN _ TIME
------------------------------------------------------
------------------------
CREATE _ SCN TOTAL _ SIZE
---------- ----------
2 PDB$SEED
10-DEC-15 12.50.37.124 PM +02:00
1594403 775946240
3 PDBORCL
10-DEC-15 12.45.51.424 PM +02:00
1608165 2148270080
CON _ ID NAME
---------- ------------------------------
OPEN _ TIME
```

```
------------------------------------------------------
-------------------------
CREATE _ SCN TOTAL _ SIZE
---------- ----------
```

13. Show Oracle service with below command.
```
SQL> select sid, username, program
from v$session
where con _ id = 0; 2 3
SID USERNAME
---------- ------------------------------
PROGRAM
--------------------------------------------------
2
oracle@rac1.localdomain (PMON)
3
oracle@rac1.localdomain (PSP0)
4
oracle@rac1.localdomain (VKTM)
SID USERNAME
---------- ------------------------------
PROGRAM
--------------------------------------------------
5
oracle@rac1.localdomain (GEN0)
6
oracle@rac1.localdomain (MMAN)
8
oracle@rac1.localdomain (DIAG)
34 rows selected.
```

14. Show connected session for a database.
```
SQL> select sid, username, program
from v$session
where con _ id = 1 2 3
4 ;
SID USERNAME
---------- ------------------------------
PROGRAM
--------------------------------------------------
53 SYS
sqlplus@rac1.localdomain (TNS V1-V3)
```

```
        69 SYS
        sqlplus@rac1.localdomain (TNS V1-V3)
```

15. Create a pluggable clone database with this command.

```
    SQL> CREATE PLUGGABLE DATABASE pdborcl1 FROM
        pdborcl
    FILE _ NAME _ CONVERT = ('/u01/app/oracle/ora-
        data/orcl/pdborcl', '/u01/app/oracle/oradata/
        orcl/pdborcl1')
    PATH _ PREFIX = '/u01/app/oracle/oradata/orcl/
        pdborcl1'; 2 3
    Pluggable database created.
```

16. Create a database with below query. (Note that whenever you enter the same name for a data file, you get an error.)

```
    Create Database
    SQL> CREATE PLUGGABLE DATABASE pdborcl4
    ADMIN USER oracle IDENTIFIED BY oracle
    ROLES = (dba)
    DEFAULT TABLESPACE users
    DATAFILE '/u01/app/oracle/oradata/orcl/pdborcl4/
        pdborcl401.dbf' SIZE 250M AUTOEXTEND ON
    FILE _ NAME _ CONVERT = ('/u01/app/oracle/oradata/
        orcl/pdbseed',
    '/u01/app/oracle/oradata/orcl/pdborcl4')
    STORAGE (MAXSIZE 2G)
    PATH _ PREFIX = '/u01/app/oracle/oradata/orcl/
        pdborcl4'; 2 3 4 5 6 7 8 9
    Pluggable database created.
```

16.1. Create the PDB open mode with below query.

```
      SQL> alter pluggable database all open;
      Pluggable database altered.
      SQL> SELECT NAME, OPEN _ MODE, RESTRICTED,
        OPEN _ TIME FROM V$PDBS;
      NAME OPEN _ MODE RES
      ------------------------------ ---------- ---
      OPEN _ TIME
      ------------------------------------------------
      ------------------------------
      PDB$SEED READ ONLY NO
      15-DEC-15 03.26.27.217 PM +02:00
      PDBORCL READ WRITE NO
```

```
15-DEC-15 03.26.36.802 PM +02:00
PDBORCL1 READ WRITE NO
15-DEC-15 03.26.36.805 PM +02:00
NAME OPEN _ MODE RES
------------------------------ ---------- ---

OPEN _ TIME
--------------------------------------------------
  ---------------------------
PDBORCL2 MOUNTED
PDBORCL3 MOUNTED
PDBORCL4 READ WRITE NO
15-DEC-15 03.26.36.800 PM +02:00
6 rows selected.
```

16.2. Add tnsnames pdb4 to tnsnames file with below query.

```
PDBORCL4 =
(DESCRIPTION =
(ADDRESS = (PROTOCOL = TCP)(HOST = localhost)
  (PORT = 1521))
(CONNECT _ DATA =
(SERVER = DEDICATED)
(SERVICE _ NAME = PDBORCL4.localdomain)
)
)
```

16.3. Finally connect to database through sqlplus.

```
SQL> conn oracle/oracle@PDBORCL4
Connected.
```

SUMMARY

As we know, cloud is trending and database management systems are moving toward this technology in which many databases are managed through a CDB. The main advantage of this technology is its consolidation for hosting many databases at a single point and managing resources easily. This chapter discusses the structures of a private cloud and PDB.

What's Exadata?

4

Oracle Corporation adopted cloud technology and developed Oracle Exadata as a new-generation relational database management system (RDBMS). As Oracle wants to keep close to a new-generation features; hence, Oracle Exadata was developed with many database administrator (DBA) and developer features. It also supports big data, and many features have been added to Oracle Exadata X7 for developing new projects that support many different applications. Today, better performance is the main aspect in solving high-complexity input/output (I/O) problems that arise in warehouse projects. Oracle Exadata provides better performance and scalability and enables easy integration of the system. Oracle Exadata shows great performance in creating great ecosystems. Oracle Exadata is the best database management system in the world, and Oracle and other vendors solve complex database problems using this system which works with compute nodes, storage nodes, and switches. It can be sized, tuned, and modified for any needs. It shows great performance even under heavy workloads, reduces database storage requirements, uses existing Oracle resources and skill sets, and results in a faster time to market. It is effective in building a new ecosystem for delivering good IT services and is a gateway to building a new IT world.

Oracle Exadata shows great performance, is cost-effective, and is applicable to Oracle database. Exadata offers private cloud features to consolidate all of the workload in sectors like banking, retail, and others. It improves performance and lowers storage costs by reducing the size of data warehousing tables by up to 10× and archive tables by up to 50×. In fact that power is needed to execute hard Structured Query Language (SQL) and online transactional processing (OLTP), particularly for warehouse systems now that Oracle Exadata CPUs are powered by Intel so IT has a high latency. Oracle develops new-generation hardware for providing Exadata service on cloud; thus, related Exadata are built with top-notch features in terms of CPU, RAM, etc. Oracle offers two different versions such as X5 and X6; thus, the service can be selected easily, and customers can continue to use a version they are familiar with. It means that if a customer uses X5 version on-premise and they don't want to change it, then they can select this version; higher versions can always be selected based on their preference [35–37,68].

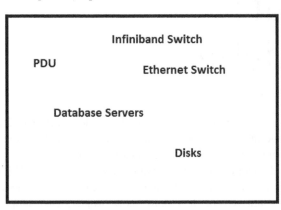

EXADATA FEATURES

---

1. **Extreme Flash Storage**
   Oracle Exadata provides extreme flash storage for all transactions and queries.
2. **Elastic Configurations**
   You can use your own customized database machines and select all products according to your needs.
3. **Engineered for the Standard Oracle Database**
   Customer applications use Oracle Database; thus, the data can be migrated for use in Oracle Exadata with no changes.
4. **Smart Scan Technology**
   Oracle Exadata provides improved query performance by offloading query processing.
5. **Oracle Exadata Smart Flash Cache**
   Oracle Exadata provides improved query response time and throughput with solid-state storage.
6. **Oracle Exadata Hybrid Columnar Compression**
   Oracle Exadata can compress numeric data by forty times [35,36,37].

SUMMARY

Oracle Exadata is a solution package that helps to improve performance, scalability, and availability, and it includes items such as InfiniBand switch, Power Distribution Unit (PDU), Ethernet switch, disks, and database servers. This chapter discusses the structure of Exadata and its features.

Oracle 12c's New Features

5

As it is known, new features of Oracle 12c were added to Oracle 12.1.0.2. After the release of Oracle 12.1.0.1, many new features for database administrator (DBA) and particularly database developer were developed. The main feature offered by Oracle is its multitenant architecture. This feature makes it easy for managing many databases through one database. Especially, Oracle is following the consolidation trend in IT. Through this feature, Oracle database easily controls consolidation of all I/O utilizations. In other words, Oracle added new features for the developer in Oracle 12.1.0.2, particularly the most expected Oracle identity column which supports to specific columns. Oracle developer can easily add an identity column with a few commands. Another important feature is the invisible column, which makes it possible for the developer to hide a column in a table, especially when companies need to hide the salary column. The heat map feature helps the DBA to manage storage and CPU usage. Therefore, Oracle DBA can arrange data which is hot or cold. The most popular feature is in-memory, and Oracle's in-memory is 40 times faster than that of the others.

This chapter aims to educate customers about the new features of Oracle 12c. It, however, addresses only the main features. For understanding the other new features of Oracle 12c, refer to Refs. [30,32,33,34].

INTRODUCTION TO ORACLE 12c's FEATURES

Heat Map and Automatic Data Optimization (ADO)

This feature provides policy-based compression and serves to track the status of data used in the system. Thus, the data that is used less can be compressed more as per the policies. Oracle's heat map feature doesn't work when it is accessed. Take a look at how to configure this feature below [31].

```
SQL> ALTER SESSION SET HEAT_MAP = ON;
Session altered.
create table system.test
(no NUMBER(22) NOT NULL,
name VARCHAR(255),
surname VARCHAR(255))
tablespace system
/
alter table system.test
ILM ADD POLICY
ROW STORE COMPRESS ADVANCED
ROW
AFTER 3 DAYS OF NO MODIFICATION
/
alter table system.test
ILM ADD POLICY
COLUMN STORE COMPRESS FOR QUERY
HIGH SEGMENT
AFTER 7 DAYS OF NO MODIFICATION /
alter table system.test
ILM ADD POLICY
COLUMN STORE COMPRESS FOR ARCHIVE
HIGH SEGMENT
AFTER 30 DAYS OF NO MODIFICATION
/
SQL> sho parameter heat;
NAME TYPE VALUE
----------------------------------------------------------------------------
heat_map string ON
```

Point Time Recovery

This feature helps to recover data from a specific date. Even if data is deleted, it can be recovered using this feature. Take a look at how to configure this feature below.

```
CONN / AS SYSDBA
CREATE USER c##user1 IDENTIFIED BY user1 QUOTA UNLIMITED
    ON users;
GRANT CREATE SESSION, CREATE TABLE TO c##user1;
CONN c##user1/user1
```

```
CREATE TABLE book (name VARCHAR2(100));
INSERT INTO book VALUES (32);
COMMIT;
SQL> SELECT DBMS_FLASHBACK.get_system_change_number FROM
    dual;
GET_SYSTEM_CHANGE_NUMBER
----------------------
1913482
INSERT INTO book VALUES (42);
COMMIT;
SQL> SELECT * FROM book;
NAME
----------------------------------------------------------------------------
32
42
[oracle@rac1 bin]$ ./rman target=/
Recovery Manager: Release 12.1.0.2.0 - Production on Tue
Dec 15 21:23:16 2015
Copyright (c) 1982, 2014, Oracle and/or its affiliates.
    All rights reserved.
connected to target database: ORCL (DBID=1425867235)
RECOVER TABLE 'c##user1'.'book'
UNTIL SCN 1913482
AUXILIARY DESTINATION '/u01/export'
REMAP TABLE 'TEST'.'book':'book_PREV';
SQL> SELECT * FROM book;
NAME
----------------------------------------------------------------------------
32
```

Identity Columns

This feature serves to increment by one the values in the table columns in an Oracle database. Take a look at how to use this feature below.

```
CREATE TABLE table1 (id NUMBER GENERATED AS IDENTITY,
    Name varchar(255));
insert into table1(Name) values ('qq');
insert into table1(Name) values ('qq1');
SQL> select * from table1;
ID
----------
NAME
----------------------------------------------------------------------------
1
```

```
qq
2
qq1
```

Invisible Columns

This feature helps to keep a column hidden in the table so that certain data are invisible. Take a look at this feature below.

```
SQL> CREATE TABLE table2 (name INT, password INT
    INVISIBLE);
Table created.
SQL> insert into table2(name,password) values('12','34');
1 row created.
SQL> select * from table2;
NAME
----------
12
```

In-Memory

This feature helps to store tables in the in-memory, and in-memory works in the system global area (SGA). Take a look at this feature below.

```
SQL> CREATE TABLE sys.table_test (
no NUMBER(5) PRIMARY KEY,
name VARCHAR2(255) NOT NULL
)
TABLESPACE tbs_test
STORAGE ( INITIAL 50K); 2 3 4 5 6
Table created.
SQL> alter table table_test INMEMORY NO INMEMORY(name);
Table altered.
SQL> CREATE TABLE tbs_test1
( q1 NUMBER,
q3 VARCHAR2(10),
q4 VARCHAR2(10) )
INMEMORY MEMCOMPRESS FOR QUERY
NO INMEMORY(q4); 2 3 4 5 6
Table created.
```

Invisible Column

In the previous version, columns cannot be hidden and made invisible. Hence, Oracle offers a new feature named invisible column, by which an Oracle developer can easily hide a column using a few commands in the query. The specific commands for using this feature are as follows:

```
SQL> CREATE TABLE table2 (name INT, password INT
    INVISIBLE);
Table created.
SQL> insert into table2(name,password) values('12','34');
1 row created.
SQL> select * from table2;
NAME
----------
12
```

Identity Column

Identity column is another feature added based on expectations of Oracle developers. In Oracle 12c, an Oracle developer can add an identity column with a few commands, an example of which is shown below:

```
CREATE TABLE table1 (id NUMBER GENERATED AS IDENTITY,
    Name varchar(255));

insert into table1(Name) values ('qq');
insert into table1(Name) values ('qq1');
SQL> select * from table1;
ID ----------
NAME
--------------------------------------------------------------------------
1 qq
2 qq1
```

Multiple Indexes

Multiple indexes weren't supported earlier but were later supported through Oracle 12c. Therefore, after adding an identity column, there is no need for more sequences and triggers, and only one command needs to be added to the "create table" query.

```
SQL> create table mindex (test1 varchar(255), test2
    varchar(255), test3 varchar(255), test4 varchar(255),
    test5 varchar(255));
Table created.
SQL> create index mindex_index on mindex(test4,test5);
Index created.
SQL> create bitmap index mindex2 on mindex(test4,test5)
    invisible;
Index created.
SQL> alter session set optimizer_use_invisible_indexes=true;
Session altered.
SQL> exec dbms_stats.set_table_stats ( user, 'mindex',
    numrows => 1000000, numblks => 100000 );
PL/SQL procedure successfully completed.
SQL> set autotrace traceonly explain
SQL>
SQL> select count(*) from mindex ;
Execution Plan
----------------------------------------------------------
Plan hash value: 537725359
----------------------------------------------------------
| Id | Operation | Name | Rows | Cost (%CPU)|
----------------------------------------------------------
| 0 | SELECT STATEMENT |  | 1 | 0 (0)|
| 1 | SORT AGGREGATE |  | 1 |  |
| 2 | BITMAP CONVERSION COUNT |  | 1000K|  |
| 3 | BITMAP INDEX FAST FULL SCAN| MINDEX2 |  |  |
----------------------------------------------------------
```

Invoker Rights

Invoker Rights is a kind of permission. Any procedure is permitted with this feature. So, a command (see "authid current_user") is added in a function that works smoothly with Oracle 12c.

Partial Indexes

Partitions are enhanced partial indexing, which can be used or not easily.

```
create table pindex
( test1 varchar(255),
test2 varchar(255),
test3 varchar(255),
```

```
test4 varchar(255),
test5 varchar(255)
)
partition by range (test1)
(partition t1 values less than (100) indexing on,
partition t2 values less than (200) indexing off
); 2 3 4 5 6 7 8 9 10 11
Table created.
SQL> begin
dbms_stats.set_table_stats
( user, 'pindex', numrows=> 1000000, numblks => 100000 );
end;
/ 2 3 4 5
SQL> select partition_name, high_value, indexing
from user_tab_partitions
where table_name = 'PINDEX'; 2 3
PARTITION_NAME
-----------------------------------------------------------------------------
HIGH_VALUE
-----------------------------------------------------------------------------
INDE
----
T1
'100'
ON
T2
'200'
OFF
PARTITION_NAME
-----------------------------------------------------------------------------
HIGH_VALUE
-----------------------------------------------------------------------------
INDE
----
PL/SQL procedure successfully completed.
```

Online Operations

Online operations are features added for developers, which support the use of
some alter and drop commands, such as "Drop Index Online" or "Alter Index
Unusable Online."

```
SQL> CREATE USER c##user IDENTIFIED BY 123
2 ;
User created.
```

```
SQL> grant dba to c##user;
Grant succeeded.
SQL> grant create session to c##user;
Grant succeeded.
SQL> grant resource to c##user;
Grant succeeded.
SQL> conn c##user/123
Connected.
SQL>
SQL> create table qqq
( test1 int,
test2 int,
test3 int
)
/ 2 3 4 5 6
Table created.
SQL> insert into qqq values ( 11 ,22 ,33 );
1 row created.
SQL> commit;
Commit complete.
SQL> update qqq set test1 = 42;
1 row updated.
SQL> select * from qqq;
TEST1 TEST2 TEST3
---------- ---------- ----------
42 22 33
SQL> declare
pragma autonomous_transaction;
begin
execute immediate 'alter table qqq set unused column
    test3 online;
end;
/ 2 3 4 5 6 ^C
SQL> declare
pragma autonomous_transaction;
begin
execute immediate 'alter table qqq set unused column
    test3 online';
end;
/ 2 3 4 5 6
PL/SQL procedure successfully completed.
SQL> select * from qqq;
TEST1 TEST2
---------- ----------
```

```
42 22
SQL> rollback;
Rollback complete.
SQL> select * from qqq;
TEST1 TEST2
---------- ----------
11 22
```

SQL Translation Framework

Migration is achieved easily with the SQL Translation Framework feature. There is no need for changing the SQL. Oracle supports Sybase, MS SQL Server, and DB2 [30].

```
-- Create a Translation Profile
SQL> exec dbms_sql_translator.create_profile('deneme');
PL/SQL procedure successfully completed.
SQL> select object_name, object_type from dba_objects
    where object_name like 'deneme';
OBJECT_NAME OBJECT_TYPE
-----------------------------------------------------------------
deneme SQL TRANSLATION PROFILE
-- Add some SQL to be translated
SQL> exec dbms_sql_translator.register_sql_
    translation('deneme','select count(*) from
    pm.employee','select count(*) from sh.employee');
PL/SQL procedure successfully completed.
You can see below side for plsql package (extend sample)
BEGIN
DBMS_SQL_TRANSLATOR.REGISTER_SQL_TRANSLATION(
profile_name => 'deneme',
sql_text => 'select count(*) from pm.employee',
translated_text => 'select count(*) from sh.employee');
END;
SQL> grant all on sql translation profile deneme to pm,sh;
Grant succeeded.
SQL> alter session set sql_translation_profile = deneme;
Session altered.
-- Enable the profile to work
SQL> alter session set events = '10601 trace name context
    forever, level 32';
Session altered.
```

Easier Reorgs

With a few commands, the tablespace can be reorganized or compressed.

```
SQL> create user c##user1 identified by 12345;
User created.
SQL> grant resource to c##user1;
Grant succeeded.
SQL> grant dba to c##user1;
Grant succeeded.
SQL> grant create session to c##user1;
Grant succeeded.
SQL> conn c##user1/12345
Connected.
SQL>
SQL>
SQL> create table TEST1
tablespace users
as
select *
from dba_users; 2 3 4 5
Table created.
SQL> begin
dbms_redefinition.redef_table
( uname => user,
tname => 'TEST1',
table_compression_type => 'row store compress advanced',
table_part_tablespace => 'USERS' );
end;
/ 2 3 4 5 6 7 8
PL/SQL procedure successfully completed.
```

SUMMARY

As we know, Oracle is the best brand for database software and technology in the world, and it is developing new products day by day. Oracle 12c seems to be its top product that has made Oracle a leader in the world's database technology industry. This chapter presents the new features of Oracle 12c.

Oracle Storage Index and Smart Scan

6

Oracle Exadata was developed as a new-generation software solution to work with engineered systems. One of its top features is storage index which works with a cell; thus, performs better in warehouse systems. Generally the smart scan feature is based on Exadata performance, but, in fact, storage is one piece of the puzzle on database performance in big warehouse projects. Many big data warehouse projects work on very large databases, and therefore, they need software to work with engineered systems like storage index which works with Oracle Exadata cells.

Storage index calculates the minimum and maximum values of specific columns. Before I/O operations and executing Where clause, Exadata calculates specific values in rows, and these values compare the column value to minimum and maximum values in storage index. If a column value is not within the range of minimum and maximum values, scan I/O for that query is avoided [69]. According to Oracle docs, this service delivers high performance in real time [38,39].

V$SYSSTAT is used for gathering related data on the effectiveness of Oracle Exadata Storage Software, and a lot of information can be stored in *V$SYSSTAT*. An example of such information is shown below [40,41].

WORKSPACE OF STORAGE INDEX

Storage index can be used with Smart Scan feature and at many points in a query. It is a special feature, and if used correctly, it can be a feature that yields very good results; perhaps one of the successes of Oracle is in implementing this feature for solutions to engineering problems.

There are some restrictions for storage index to work smoothly in a system. For instance, if storage index feature needs to execute a query, it must use a prediction filter.

P.S. As it is known, there is a specific condition for *IS NULL* [40,41,42].

ORACLE STORAGE INDEX HIDDEN PARAMETERS

As said, there are not many commands available for changing the parameters of storage index that is used in Exadata. Many hidden parameters can be shared, but as it is known, these parameters are not officially supported by Oracle; hence, be careful in using these commands as support issues may arise in future.

_KCFIS_STORAGEIDX_DISABLED

As it is known, there are not many ways to configure storage index, but some hidden parameters, such as the *_kcfis_storageidx_disabled* command, affect the system directly. So, Exadata storage index capabilities can be disabled through this command, but normally the default value is "false." Commands to configure storage index are as follows:

Enabling Storage Index

```
Alter system set _kcfis_storageidx_disabled=false.
```

Disabling Storage Index

```
Alter system set _kcfis_storageidx_disabled=true [40,41].
```

CELL_STORIDX_MODE

Status of storage index can be controlled with _cell_storidx_mode, and storage index can be set through this command that can be used with three options, namely *ALL*, *KDST*, and *EVA*, and these options can be used for different requests on Exadata; for instance, *EVA* provides support to all compression operators, but *IS NULL* isn't supported by *EVA*. Examples of this command are as follows:

```
Alter session set _cell_storidx_mode=EVA;
Alter session set _cell_storidx_mode=ALL;
Alter session set _cell_storidx_mode=KDST.
```

—

This mode is used by default. EVA is the Oracle Kernel name, and ALL applies to all compression methods.

ORACLE EXADATA SMART SCAN

Oracle Exadata was introduced by Oracle, and customers were asked to give their opinion about the performance of its features. Customers had high expectations of Oracle, and Exadata sales show that they like the product. Smart scan is the main feature of Oracle Exadata; Oracle storage software includes many features such as smart scan and storage index [38,39].

```
ALTER SESSION SET CELL_OFFLOAD_PROCESSING = FALSE;
select * from t1 where OWNER='SYSTEM';

129 rows selected.

Elapsed: 00:00:01.69

Execution Plan
```

```
----------------------------------------------------------
Plan hash value: 3617692013


---------------------------------------------------------------------
| Id  | Operation                    | Name | Rows  | Bytes
| Cost (%CPU)| Time
---------------------------------------------------------------------
|   0 | SELECT STATEMENT             |      |   114 | 34998
|   158   (0)| 00:00:01
|*  1 |   TABLE ACCESS STORAGE FULL| T1   |   114 | 34998
|   158   (0)| 00:00:01
---------------------------------------------------------------------


Predicate Information (identified by operation id):
---------------------------------------------------

   1 - storage("OWNER"='SYSTEM')
       filter("OWNER"='SYSTEM')

ALTER SESSION SET CELL_OFFLOAD_PROCESSING = TRUE;

select * from t1 where OWNER='SYSTEM';

129 rows selected.

Elapsed: 00:00:01.58

Execution Plan
----------------------------------------------------------
Plan hash value: 3617692013
---------------------------------------------------------------------
| Id  | Operation                    | Name | Rows  | Bytes
| Cost (%CPU)| Time
---------------------------------------------------------------------
|   0 | SELECT STATEMENT             |      |   114 | 34998
|   158   (0)| 00:00:01
|
|*  1 |   TABLE ACCESS STORAGE FULL| T1   |   114 | 34998
|   158   (0)| 00:00:01
---------------------------------------------------------------------
--


Predicate Information (identified by operation id):
---------------------------------------------------
```

DIRECT PATH READ

In fact, power is needed to execute hard structured query language in warehouse systems and online transactional processing (OLTP). Particularly in warehouse systems, as we know, Oracle Exadata is powered by Intel. Oracle uses a new-generation hardware for providing Exadata service on cloud; thus, related Exadata are provided with top-notch features in terms of CPU and RAM. The main working principle of Exadata's smart scan is direct reading from the path due to performance bottlenecks. It means that Direct Path Read enables us to bypass the buffer cache and directly read from the related block; in other words, this technology enables us to skip a dirty block (A dirty block is a block that does not have the same image in memory and on disk [70]), and thus, it can be used by the DBA [43].

Some delays can be encountered with Direct Path Read, but they could be caused due to various reasons.

Direct Path Read can be enabled and disabled as shown below.

The status of _serial_direct_read_ can be understood with a view and explained as follows:

```
SQL> select KSPPSTVL from x$ksppcv join x$ksppi using
    (indx) where ksppinm='_serial_direct_read';

KSPPSTVL
----------------------------------------------------------------------------
auto
```

P.S. Please execute above query at SYS user; otherwise the query will not be executed due to authorization issues.

Normal decision support systems (DSS) systems are configured with "AUTO," but enabling or disabling can directly be configured by the user as follows:

Enabling Direct Path Read

```
Alter session set _serial_direct_read=true.
```

Disabling Direct Path Read

```
Alter session set _serial_direct_read=false.
```
_serial_direct_read_ affects other executions of query particularly IAS (Insert as Select). It means whenever a query is executed with full hint and the system is configured "true" in _serial_direct_read_, the query will be executed

faster because the execution is supported by smart scan. By the way, for a better undertanding of the effects of _serial_direct_read on IAS, look at the My Oracle Support (MOS) [ID 1348116.1] note.

Let's take a look at the effect of _serial_direct_read on IAS.

First take a look at the status of _serial_direct_read.

```
SQL> select KSPPSTVL from x$ksppcv join x$ksppi using
     (indx) where ksppinm='_serial_direct_read';

KSPPSTVL

--------------------------------------------------------------------------------
Auto
Create table was created  as below .
SQL> create table t1 as select * from dba_tables;
Table created.
Elapsed: 00:00:01.16
Configuration was set as "false" ...
SQL> alter session set "_serial_direct_read"=false;
Session altered.
Now, execute IAS as follows and observe the execution
time.

SQL> Insert into t1 select /*+full(w)*/ * from dba_tables;
4657 rows created.
```
Elapsed: 00:00:03.28
```
As shown above, after 3 seconds, observe with the other
   configuration.
SQL> alter session set "_serial_direct_read"=true;
Session altered.
Now, execute IAS as follows, and observe the execution
time.
SQL> Insert into t1 select /*+full(w)*/ * from
dba_tables;
4657 rows created.
```
Elapsed: 00:00:00.33
```
As shown above, the execution is six times faster
   than the old configuration and this gives the
   user a better understanding of the effects of
   _serial_direct_read.
On the other hand, using IAS with  APPEND hint would
   result in better performance and faster execution of
   the query.
P.S. When  CTAS (Create Table as Select) is used with
   this feature, the performance may not be as good as
   that of the other configurations.
```

FULL TABLE SCAN

As it is known, full table scan reads all rows of the table and looks for other criteria that do not match. On the other hand, the optimizer selects full table scan when the user does not select the other pathway. Full table scan reads the database sequentially one block at a time. Oracle Exadata smart scan is designed for better performance; it executes a query with full table scan [40,41,43].

Let's look at a full table scan on Exadata.

```
Elapsed: 00:00:00.08

Execution Plan
----------------------------------------------------------------
Plan hash value: 3617692013
----------------------------------------------------------------
| Id  | Operation                  | Name | Rows  | Bytes
| Cost (%CPU)
|
Time--------------------------------------------------------------
-----------------

|   0 | SELECT STATEMENT           |      |       |  4656 |
100K|   158    (0)

| 00:00:01|  1 | TABLE ACCESS STORAGE FULL| T1   |
4656 |   100K|   158    (0)
| 00:00:01
----------------------------------------------------------------
------------
    -----
```

In-Memory Options on Smart Scan

Optimizer can be selected in the following way and tested for in-memory.

```
|   2 |    PX SEND QC (RANDOM)           | :TQ10000 |   114
| 34998 |     3   (34)|
00:00:01 |  Q1,00 | P->S | QC (RAND)  |
```

```
|   3  |      PX BLOCK ITERATOR              |        |    114
| 34998 |       3   (34) |
00:00:01 |  Q1,00 | PCWC |                   |
|*  4  |      TABLE ACCESS INMEMORY FULL| T1      |    114
| 34998 |       3   (34) |
00:00:01 |  Q1,00 | PCWP |                   |
```

CELL_OFFLOAD_DISPLAY

This command works like a mechanism; in other words, SQL EXPLAIN PLAN displays predicates for evaluating Exadata. *AUTO, ALWAYS*, and *NEVER* are some of the available [42,43].

```
ALTER SESSION SET CELL_OFFLOAD_DISPLAY = ALWAYS;
```

CELL_OFFLOAD_DECRYPTION

This command is used particularly to encrypt tablespace and column as follows:

```
ALTER SYSTEM SET CELL_OFFLOAD_DECRYPTION = TRUE.
```

SUMMARY

Oracle storage index is integrated directly. It means that there are lots of ways for configuring the system; generally, an *execute* command is used to start storage index, and after that, this feature can be controlled through ways that can be found easily on docs.oracle. This chapter explains the basics of storage index, and the later chapters will explain Exadata in more detail.

Oracle Exadata I/O Resource Manager

7

Consolidation is an acceptable solution to handle mixed workload in a company that generates small or large amount of data. Most of the companies maintain databases, application servers, third-party applications to successfully run their businesses, and this trend has increased in a short period of time. We know that resources are one of the most important inputs for a business, particularly IT business. Initially companies don't understand the usage of IT tools in their business, but when they face a problem, they are unable to locate the correct tools they require due to lack of perfect plans, which is not good for consolidating the usage of resources. Unfortunately, resources aren't unlimited; thus, companies must quickly understand the value of resources. For instance, many companies use applications on different servers, and they often do not know the number of resources they need to use for an application, which may lead to increase in IT costs. To prevent problems created by weak IT consolidation, nowadays compaines are addressing this issue [63].

Exadata machine is used for online transaction processing (OLTP) and warehouse and mixed loads. This system is one of the best database machines produced in the past decade. Many important issues such as consolidation can be addressed using Exadata. One of its most important features is its multitenant architecture which is a primary feature in Oracle 12c that allows you to manage many databases very conveniently under one roof by controlling the use of input/output. Thanks to this feature, the cost can be reduced as many databases can be managed under a single roof. For example, multiple databases can be connected to all subdatabases through a CDB. Side by side, it is very easy to create a clone database with a few clicks, thanks to this feature. Of course, this feature has other benefits as well.

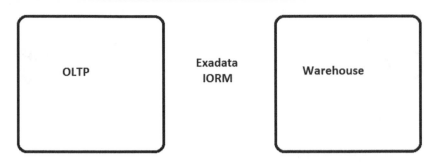

Warehouse and OLTP and many other features that support these two features are available in Exadata. Initially, I/O resource manager (IORM) was the resource management system, and this feature was available only in Oracle database before Oracle Exadata.

UNDERSTANDING IORM

Oracle Exadata database machine uses an IORM system managed by many databases. The biggest problem in a database system is inadequate resource consumption, which means that the data is output slowly from poor environments, but if the resource consumption is planned and followed properly, the data output rate can be increased very easily. IORM can be confidently configured by creating a resource plan. The IORM system is available in the Oracle Enterprise version.

IORM Plan

This command shows the status of IORMPLAN as follows:

```
CellCLI> list iormplan detail
     name:              dm01cel01_IORMPLAN
     catPlan:
     dbPlan:
     objective:         basic
     status:            active
```

Default Plan

If needed, set the default value for the plan.

```
CELLCLI> ALTER IORMPLAN dbPlan=""

CELLCLI> ALTER IORMPLAN catPlan=""
```

ALTER IORM PLAN

```
CELLCLI>ALTER IORMPLAN
((name=test_db, slevel=1, allocation=70),-
((name=test_db2, slevel=1, allocation=70),-
(name=test_db3, slevel=1, allocation=70),-
((name=test_db4, slevel=1, allocation=70),-
((name=test_db5, slevel=1, allocation=70),-
((name=test_db6, slevel=1, allocation=70);
```

ALTER IORM PLAN for Data Guard

IORM can be used for data guard, and IORM Data Guard can be used as shown below.

```
CELLCLI>ALTER IORMPLAN
((name=test_db2, share=8, role=primary),-
((name=test_db2, share=8, role=primary),-
(name=test_db3, share=1, role=standby),-
((name=test_db4, share=1, role=primary),-
((name=test_db5, share=1, role=primary),-
((name=test_db6, share=2, role=primary);
```

ALTER IORM PLAN in Smart Flash Cache

Interdatabase plan can be configured with smart flash cache feature. IORM Data Guard can be used as shown below.

```
CELLCLI>ALTER IORMPLAN
((name=test_db2, share=8, limit=50, flashlog=off),-
((name=test_db2, share=8, limit=25, flashlog=off);
```

INTERDATABASE PLAN USING ALLOCATION

Interdatabase plan can be configured as a combination of catplan and dbplan and used as shown below.

ALTER IORM PLAN

```
CELLCLI>ALTER IORMPLAN
catPlan=((name=test_db, slevel=1, allocation=70),-
       ((name=test_db2, slevel=1, allocation=70),-
dbplan=(name=test_db3, slevel=1, allocation=70),-
       ((name=test_db4, slevel=1, allocation=70),-
       ((name=test_db5, slevel=1, allocation=70),-
       ((name=default ,share=2);
```

SUMMARY

"Consolidation" is a keyword in the IT industry, and Oracle makes the best use of this concept. Exadata is a machine built entirely for consolidation and is very easy to operate and manage. This chapter discusses the resource management between OLTP and DSS warehouse.

Hybrid Columnar Compression (HCC)

8

Nowadays, Oracle uses Exadata, as it provides extreme performance; thus, Oracle would want to protect this reputation. It can be accepted that today there are two things needed for being better in terms of technology: more I/O and more space in your system. I/O can be generated with extreme performance machines like Exadata and Sparc, but how to save more disk space in our system and thus prevent spending more for purchasing new disks? The graphs related to the usage of a disk show a dramatic increase of data; as the growth of data cannot be stopped, new disks are needed for accomodating more data. IT organizations need new disks due to increase of data worldwide. The method of storing data has changed in recent years, and thus only some cheap disks are chosen, especially for cloud storage, but if data continues to increase in this way in the coming years, disks will not be enough for storing data; therefore, storage vendors are tending toward developing a new technology for compression. In getting started with the new-generation disk technology for saving money and time, many new features have been developed as mentioned. Hybrid columnar compression (HCC) is one of the best features of Exadata which is used to help customers save more disk space through new-generation compression.

When Oracle Exadata was proudly released, some features like HCC were expected. This technology helps to compress data in the latest Exadata and compresses data better than the normal compression technology. It can compress data up to 40 times; therefore, data can't occupy more space in the disk. More can be said about HCC that is used generally for numeric data like telco data. HCC is a database feature related to storage. Saving of storage space can be greatly increased using HCC. When data is stored and compressed together, it is stored in a compression unit [44,45].

Compression Unit

Block Header	
Compression Unit Header	
C1	C2

HCC can provide high storage space saving, but data must be loaded using the following method of data warehouse bulk loading [46].

1. Create Table as a Select (CTAS)
2. Insert statement can be used with APPEND hint
3. Direct Path Loader SQL*LDR

Some compression methods are explained below.

COMPRESSION METHOD	SYNTAX	DESCRIPTION
Basic table compression	ROW STORE (BASIC)	Insert and update rows are uncompressed.
Advance row compression	ROW STORE (ADVANCED)	Insert and update rows are compressed.
Warehouse compression (HCC)	COMPRESS FOR QUERY (LOW/HIGH)	Update rows are stored in row format.
Archive compression (HCC)	COMPRESS FOR ARCHIVE (LOW/HIGH)	Update rows are stored in row format.

BASIC TABLE COMPRESSION

```
CREATE TABLE t2_basic_c ROW STORE COMPRESS BASIC as
    select * from t2;

select * from t2_basic_c;
```

```
0,202 47MB

Alter table t2_basic_c nocompress;

select * from t2_basic_c;
0,024 47MB

SELECT SEGMENT_NAME, BYTES/1024/1024 MB FROM USER_
  SEGMENTS WHERE SEGMENT_NAME LIKE 'T2_BASIC_C';
```

ADVANCE ROW COMPRESSION

```
CREATE TABLE t2_advance_c ROW STORE COMPRESS ADVANCED as
select * from t2;

select * from t2_advance_c;
0,344   52MB

Alter table t2_basic_c nocompress;

select * from t2_advance_c;
0,135   52MB
```

WAREHOUSE COMPRESSION

Query Low Compression

```
CREATE TABLE t2_ql_c COMPRESS FOR QUERY LOW as select *
from t2;

select * from t2_ql_c;
0,378 22MB

Alter table t2_ql_c nocompress;

select * from t2_ql_c;
0,105 22MB

SELECT SEGMENT_NAME, BYTES/1024/1024 MB FROM USER_
  SEGMENTS WHERE SEGMENT_NAME LIKE 'T2_QL_C';
```

Query Low High

```
CREATE TABLE t2_qh_c COMPRESS FOR QUERY HIGH as select *
  from t2;

select * from t2_qh_c;
0,331 13MB

Alter table t2_qh_c nocompress;

select * from t2_qh_c;
0,135 13MB

SELECT SEGMENT_NAME, BYTES/1024/1024 MB FROM USER_
  SEGMENTS WHERE SEGMENT_NAME LIKE 'T2_QH_C';
```

ARCHIVE COMPRESSION

Archive Low Compression

```
CREATE TABLE t2_al_c COMPRESS FOR ARCHIVE LOW as select *
  from t2;

select * from t2_al_c;
0,215 7MB

Alter table t2_al_c nocompress;

select * from t2_al_c;
0,031 7MB

SELECT SEGMENT_NAME, BYTES/1024/1024 MB FROM USER_
  SEGMENTS WHERE SEGMENT_NAME LIKE 'T2_AL_C';
```

Archive High Compression

```
CREATE TABLE t2_ah_c COMPRESS FOR ARCHIVE HIGH as select
* from t2;

select * from t2_ah_c;
```

```
0,47 5MB

Alter table t2_ah_c nocompress;

select * from t2_ah_c;
0,187 5MB

SELECT SEGMENT_NAME, BYTES/1024/1024 MB FROM USER_
  SEGMENTS WHERE SEGMENT_NAME LIKE 'T2_AH_C';
```

12	T2	502
13	T2_ADVANCE_C	52
14	T2_AH_C	5
15	T2_AL_C	7
16	T2_QH_C	13
17	T2_QL_C	22

DSBMS

Status of table compression
```
SELECT table_name, compression, compress_for FROM
  user_tables;
```

Status of partition related to table compression
```
SELECT table_name, partition_name, compress_for
  FROM user_tab_partitions;
```

SUMMARY

The IT industry knows cost is a big factor in the world and storage space must be saved with the help of the features of the system. Exadata can save storage space with the HCC feature that compresses data 40 times, and this chapter explains this feature for a better understanding.

CellCLI

9

As it is known, Oracle Exadata focused on storage technology to achieve high performance and easy management. Technologies are integrated to focus directly on storage and storage-related features like smart scan and storage index. In fact, new-generation technologies are developing new storage methods for achieving better results in mixed workload conditions, but this is not an easy practice with vintage technology. However, many companies still haven't changed their idea of using old technology to avoid high costs, but some companies are changing this idea and seeking new-generation storage methods, such as using full SSD in storage or new-generation software in a machine like Oracle Exadata which is one of the best examples for this case. Actually, customers would want to see not only performance on storage but also affordable price Oracle Exadata is one of the top machines that show extreme performance, and for this performance, smart scan or storage index technology is used [47–49].

Let's start with the enhancement of storage management, namely disk management, which is one of the chaos cases. Earlier, the system couldn't be managed well as we could rarely take support from vendors, but today this case has changed, and the whole system can be managed through command line, web interface, and application programming interface (API). Oracle storage server is an optimized solution that works with Oracle storage software to store and access Oracle database. Oracle storage software is designed to achieve maximum performance in executing high-complexity queries, but there are some rules that need to be followed for this [50–52].

According to storage technology that is powered by Exadata CELL technology, namely all of management are to configure although CELL system that can be used over the CellCLI (cell control command-line interface) command.

Some of the capabilities of the CellCLI are as follows [53–55]:

1. ALERT
2. DROP
3. CREATE

4. GRANT
5. LIST
6. REVOKE
7. IMPORT CELLDISK
8. EXPORT CELLDISK.

ALERTHISTORY	ALERTHISTORY shows the list for alerts.
ALERTDEFINITION	ALERTDEFINITION shows the explanation for every alert.
CELL	CELL shows information on cells such as this command "list cell"[71].
GRIDDISK	GRIDDISK shows a logical partition of cell disk.
PHYSICALDISK	PHYSICALDISK shows the disks available on a cell.

Cell disks can be listed with all details using the following script.

```
CellCLI> list cell detail
      name:              dm01cel01
      accessLevelPerm:       remoteLoginEnabled
      cellVersion:       OSS_12.2.1.1.1_LINUX.
                           X64_170419
      cpuCount:          32/32
      diagHistoryDays:     7
      fanCount:          8/8
      fanStatus:         normal
      flashCacheMode:      WriteBack
      id:            1506NM703P
      interconnectCount: 2
      interconnect1:     ib0
      interconnect2:     ib1
      iormBoost:         0.0
```

Grid disks can be listed with their disk status as follows:

```
CellCLI> list griddisk
      DATA_FD_00_dm01cel01        active
      DATA_FD_01_dm01cel01        active
      DATA_FD_02_dm01cel01        active
      DATA_FD_03_dm01cel01        active
      DATA_FD_04_dm01cel01        active
      DATA_FD_05_dm01cel01        active
      DATA_FD_06_dm01cel01        active
      DATA_FD_07_dm01cel01        active
      DBFS_DG_FD_02_dm01cel01        active
```

This command is used to create a cell disk, and there are some alternatives for creating these disks on a cell.

```
CellCLI> CREATE CELLDISK ALL
```

This command is used to create a cell disk and add prefix so that it can be scattered as needed such as data or reco.

```
CellCLI> CREATE GRIDDISK ALL HARDDISK PREFIX=data,
    size=200G
```

Other uses are as follows:

```
CellCLI> CREATE GRIDDISK ALL HARDDISK PREFIX=reco,
    size=500G
```

The sparse grid disk uses up to 300 GB of the physical cell disk size, but it exposes 20,000 GB of virtual space for the Oracle automatic storage management (ASM) files.

```
CellCLI> CREATE GRIDDISK ALL HARDDISK PREFIX=sp,
    size=300G, virtualsize=20000G
```

Oracle Exadata has cell servers that are capable of creating flash cache, and before creating flash cache, its size must be set.

```
CellCLI> CREATE FLASHCACHE ALL size=100g
Flash cache cell01_FLASHCACHE successfully created
```

The details of flash cache are listed below.

```
CellCLI> LIST FLASHCACHE DETAIL
    name:              dm01cel01_FLASHCACHE
    cellDisk:
FD_00_dm01cel01,FD_05_dm01cel01,FD_03_dm01cel01,FD_07_
    dm01cel01,FD_06_dm01cel01,FD_01_dm01cel01,FD_04_
    dm01cel01,FD_02_dm01cel01
    creationTime:    2017-06-19T14:55:
                            40+03:00
    degradedCelldisks:
    effectiveCacheSize:   596.125G
    id:              cd58af9d-a55b-47b1-91bb-
                        75e403ef9598
    size:            596.125G
    status:          normal
```

DROP FLASHCACHE

Flash cache can be removed as shown below.
```
CellCLI> DROP FLASHCACHE
```

CURRENT METRIC

Exadata uses the current metric for displaying the time.

```
CellCLI> LIST METRICCURRENT WHERE objectType = 'CELLDISK'
    CD_BY_FC_DIRTY        FD_00_dm01cel01      0.000 MB
    CD_BY_FC_DIRTY        FD_01_dm01cel01      0.000 MB
    CD_BY_FC_DIRTY        FD_02_dm01cel01      0.000 MB
    CD_BY_FC_DIRTY        FD_03_dm01cel01      0.000 MB
    CD_BY_FC_DIRTY        FD_04_dm01cel01      0.000 MB
    CD_BY_FC_DIRTY        FD_05_dm01cel01      0.000 MB
    CD_BY_FC_DIRTY        FD_06_dm01cel01      0.000 MB
    CD_BY_FC_DIRTY        FD_07_dm01cel01      0.000 MB
    CD_IO_BY_R_LG         FD_00_dm01cel01      6,144,546 MB
    CD_IO_BY_R_LG         FD_01_dm01cel01      6,186,089 MB
    CD_IO_BY_R_LG         FD_02_dm01cel01      6,150,969 MB
    CD_IO_BY_R_LG         FD_03_dm01cel01      6,196,292 MB
    CD_IO_BY_R_LG         FD_04_dm01cel01      6,177,780 MB
    CD_IO_BY_R_LG         FD_05_dm01cel01      6,184,511 MB
    CD_IO_BY_R_LG         FD_06_dm01cel01      6,233,781 MB
    CD_IO_BY_R_LG         FD_07_dm01cel01      6,137,112 MB
    CD_IO_BY_R_LG_SEC     FD_00_dm01cel01      0.000 MB/sec
    CD_IO_BY_R_LG_SEC     FD_01_dm01cel01      0.000 MB/sec
    CD_IO_BY_R_LG_SEC     FD_02_dm01cel01      0.000 MB/sec
```

METRIC HISTORY

With the LIST METRIC HISTORY command, the metric history for a cell can be listed as shown below.

```
CellCLI> LIST METRICHISTORY CD_IO_RQ_R_LG
     CD_IO_RQ_R_LG     FD_00_dm01cel01          36,705,580
IO requests  2017-09-27T00:56:05+03:00
     CD_IO_RQ_R_LG     FD_01_dm01cel01          36,811,884
IO requests  2017-09-27T00:56:05+03:00
     CD_IO_RQ_R_LG     FD_02_dm01cel01          36,637,516
IO requests  2017-09-27T00:56:05+03:00
     CD_IO_RQ_R_LG     FD_03_dm01cel01          36,955,011
IO requests  2017-09-27T00:56:05+03:00
     CD_IO_RQ_R_LG     FD_04_dm01cel01          37,100,969
IO requests  2017-09-27T00:56:05+03:00
     CD_IO_RQ_R_LG     FD_05_dm01cel01          36,790,521
IO requests  2017-09-27T00:56:05+03:00
```

SUMMARY

As known, Oracle Exadata is a machine having storage features, enable us to manage disk space. This chapter explains how to use storage space in Oracle Exadata.

EXAchk

10

A database is an important part of a system that needs attention, and hence, understanding it is a specific task for the DBA and developer. As known, no database product can work independently, not even one brand. They work with some other IT components. A database product needs an operating system, dedicated network, and specific hardware such as a server, RAM, and CPU. As known, many of these products do not work smoothly always; thus, some steps are needed to understand and prevent problems in future. This shows that a product may not be capable of solving your problem and would hence be a faulty solution for your system. Oracle is aware of this case of enterprise company database and generally invests more in its support and development teams than sales teams. Oracle has large teams for developing products and providing support to its customers; in particular, Oracle gives priority to Oracle database and products for understanding database problems and providing solutions [62].

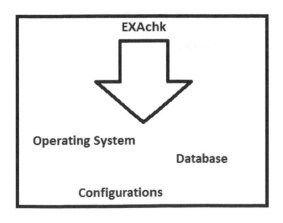

EXAchk

As explained, Oracle Exadata is one of the first database machines and also one of the most important products in the world, but as with other servers, it needs maintenance; therefore, some scripts were introduced to diagnose Oracle database and other products such as Exadata and Exalogic, which provide some information related to database and operating system using some specific operating system commands.

1. First of all, open Oracle support page for finding EXAchk script. Oracle support provides official solutions to problems related to Oracle product families that include database application servers, E-business solutions, and other products.

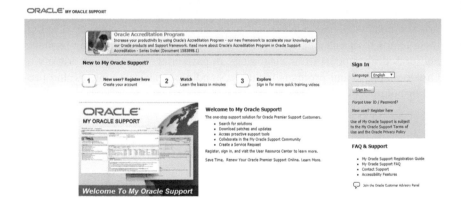

2. There are many pages related to Exadata on Oracle support page, but the following page contains related Exadata diagnostic script (Doc ID 1070954.1).

3. View the currently produced version shown below, click on the link, and download the specific files.

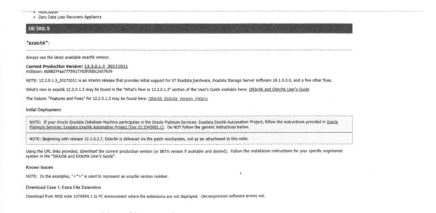

4. Now, move EXAchk.zip file to Exadata.

Name	Date modified ▾	Type	Size
exachk.zip	21.10.2017 17:01	Compressed (zippe...	19.230 KB

5. Unzip EXAchk.zip file.
   # unzip exachk.zip
6. Before executing the script, go through the readme file named as ***ORAchk_and_EXAchk_User_Guide.pdf*** for understanding the execution mechanism.

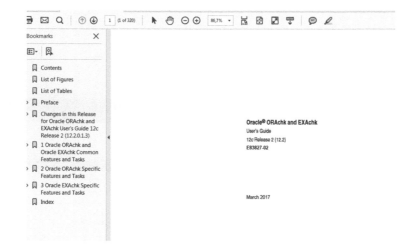

7. Go for understanding EXAchk with the following guide.

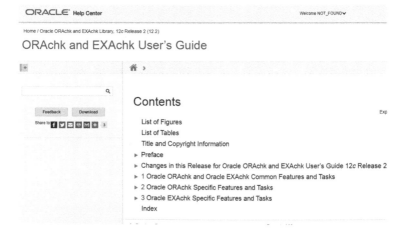

8. Control the EXAchk version with the following command.
   ```
   ./exachk   -v
   ```
9. Start EXAchk script with the following command.
   ```
   ./exachk -a
   ```
10. Select a database needed to generate a report.
11. Enter *root* password for generating a report; ensure the correctness of the password.
12. Download the files.
13. Finally the reports can be converted to HTML files.

SUMMARY

Every database may not work in a healthy manner; hence, an understand of database health is needed. Oracle understands this problem, and introduced EXAchk. This chapter explains this feature.

Database Upgrade

11

Oracle announced the release of its new database, Oracle Database 12c, in June 2013. The new release promises to deliver comprehensive features for DBAs and developers. In order to make use of these features, databases need to be upgraded to Oracle Database 12c. Oracle offers a number of upgrade options for technical persons. The choice of selecting the upgrading method can vary between companies according to their status and business criticality, database size, outage toleration, and technical expertise. These options support the updgrading from Oracle Database 11g to Oracle Database 12c. This section is designed for upgrading databases Oracle Database 12c. All of this upgrade tutorial is not tested in Exadata. All tutorial work is done on a a virtual machine. It contains all the necessary details required for the upgrade. Therefore, the attached virtual machine contains both Oracle Database 11.2.0.4 and Oracle Database 12.1.0.2 [56–58].

As stated earlier, there are some methods for upgrading. These are

1. DBUA (Database Upgrade Assistant)
2. Transportable Tablespace
3. Manual Upgrade.

DBUA (DATABASE UPGRADE ASSISTANT)

DBUA can be used by many Oracle DBAs as the upgrade process can be handled easily and finished quickly. In addition to the facts stated above, Oracle DBUA reduces the probability of wrong execution of scripts in the upgrade process and increases the ratio of successful upgrades. Therefore, it can be stated that DBUA is one of the best ways for newbie DBAs in the upgrade process [59–61].

The steps in the upgrade process could be listed as follows [64–67]:

1. Select the operation you want to perform.

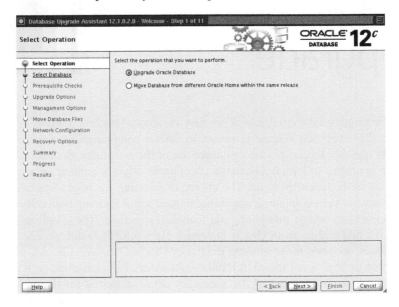

2. Select the database that you want to upgrade to on the window.

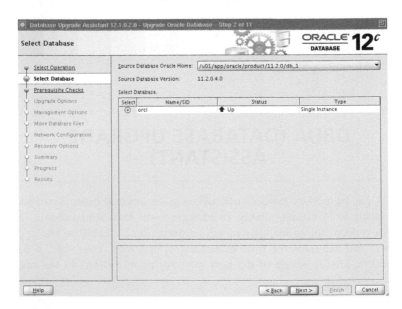

3. Now, the system checks for prerequisites before the upgrade.

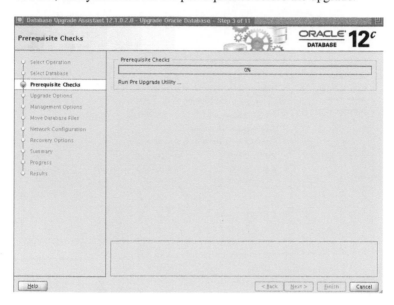

4. DBUA screen shows the result of prerequisite checks, errors, and warnings. Besides these, DBUA can generate fixup scripts for errors and warnings automatically.

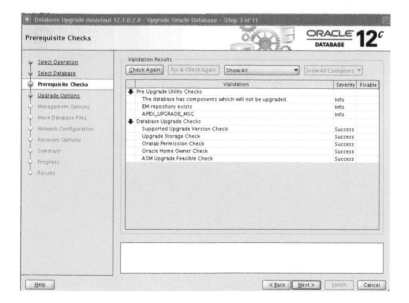

5. In this step, DBUA asks you to specify the upgrade options.

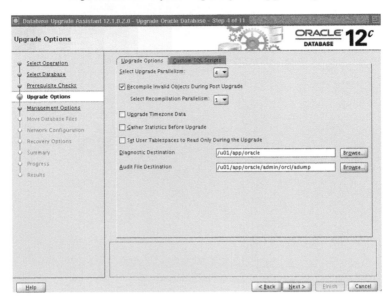

6. Enterprise Manager (EM) has to be configured in this step.

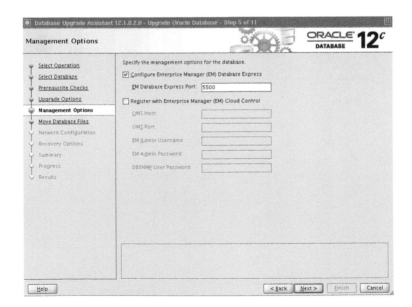

7. Some projects need the database files to be moved; hence, Oracle DBUA helps you to move your database file in this step. This step also helps to allocate FRA (Flash Recovery Area).

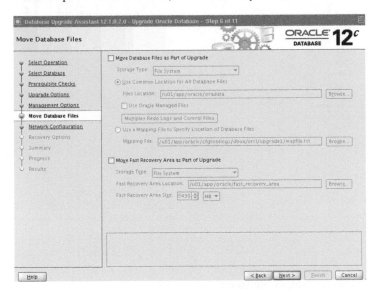

8. This step is related to Oracle Listener upgrade. In addition to this, if the database listener is not open during this process, this window can't show it.

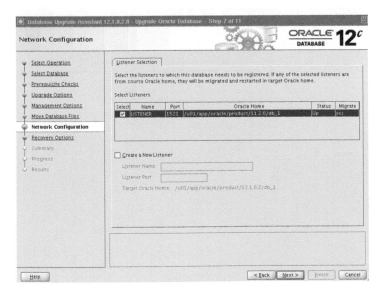

9. This step allows you to take a backup before upgrade.

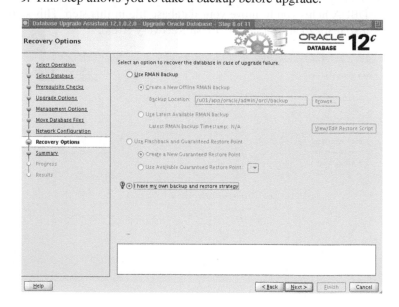

10. This step shows the summary of a smooth upgrade.
11. A window shows the progress and status of all the upgrade steps.
12. After completion of the upgrade, click "Upgrade Result."

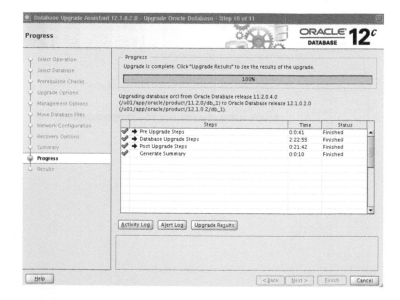

13. The Upgrade Results page displays the summary of the process.
14. Check the database upgrade results on sqlplus.

```
[oracle@upgrade ~]$ sqlplus / as sysdba
SQL*Plus: Release 12.1.0.2.0 Production on Tue
   Nov 3 03:43:02 2015
Copyright (c) 1982, 2014, Oracle. All rights
   reserved.
Connected to:
Oracle Database 12c Enterprise Edition Release
   12.1.0.2.0 - 64bit Production
With the Partitioning, OLAP, Advanced Analytics
   and Real Application Testing options
SQL>

By using DBUA, Oracle Database 11.2.0.4 has been
   upgraded to 12.1.0.2
```

TRANSPORTABLE TABLESPACE

Before the upgrade process, database configuration assistant (DBCA) is initialized.

```
[oracle@upgrade bin]$ export ORACLE_HOME=/u01/app/oracle/
   product/11.2.0/db_1/

[oracle@upgrade bin]$ PATH=$ORACLE_HOME/bin:$PATH

[oracle@upgrade bin]$dbca
```

1. First of all, environment variables must be set.

```
[oracle@upgrade bin]$ export ORACLE _ SID=orcl

[oracle@upgrade bin]$ ORAENV _ ASK=NO

[oracle@upgrade bin]$ . oraenv

The Oracle base remains unchanged with value /
   u01/app/oracle

[oracle@upgrade bin]$ ORAENV _ ASK=YES
```

2. After environment settings, the database must be connected using a command line via sqlplus.

    As an initial step in the database, a directory object user, and tablespace are created, and privileges are granted to the user. Finally, the status of tablespaces is changed to read-only mode.

```
[oracle@upgrade bin]$ sqlplus / as sysdba
SQL*Plus: Release 11.2.0.4.0 Production on Mon
  Nov 2 09:18:53 2015
Copyright (c) 1982, 2013, Oracle. All rights
  reserved.
Connected to:
Oracle Database 11g Enterprise Edition Release
  11.2.0.4.0 - 64bit Production
With the Partitioning, OLAP, Data Mining and
  Real Application Testing options
SQL> CREATE OR REPLACE DIRECTORY TEMP _ DIR AS
  '/tmp/';
Directory created.
SQL> CREATE TABLESPACE data _ ts
DATAFILE    '/u01/app/oracle/oradata/orcl/data01.
  dbf' SIZE 1M
AUTOEXTEND ON NEXT 1M; 2 3
Tablespace created.
SQL> CREATE user test IDENTIFIED BY test
DEFAULT TABLESPACE data _ ts
TEMPORARY TABLESPACE temp
QUOTA UNLIMITED ON data _ ts; 2 3 4
User created.
SQL> GRANT CREATE SESSION, CREATE TABLE TO
  test;
Grant succeeded.
SQL> CREATE TABLE test.t1 AS
SELECT * FROM dba _ objects; 2
Table created.
SQL> ALTER TABLESPACE data _ ts READ ONLY;
Tablespace altered.
SQL> ALTER TABLESPACE example READ ONLY;
Tablespace altered.
SQL>
SQL> EXIT
```

3. The system starts exporting from the database.

```
[oracle@upgrade bin]$
[oracle@upgrade      bin]$      expdp      system
full=Y    transportable=always    version=12
directory=TEMP _ DIR \
>    dumpfile=orcl.dmp    logfile=expdporcl.log
exclude=TABLESPACE:\"= \'USERS\'\"
Export: Release 11.2.0.4.0 - Production on Mon
Nov 2 09:20:31 2015
Copyright (c) 1982, 2011, Oracle and/or its
affiliates. All rights reserved.
Password:
Connected to: Oracle Database 11g Enterprise
Edition Release 11.2.0.4.0 - 64bit Production
With the Partitioning, OLAP, Data Mining and
Real Application Testing options
Starting "SYSTEM"."SYS _ EXPORT _ FULL _ 01": sys-
tem/******** full=Y transportable=always ver-
sion=12 directory=TEMP _ DIR dumpfile=orcl.dmp
logfile=expdporcl.log    exclude=TABLESPACE:"=
'USERS'"
Estimate in progress using BLOCKS method...
Processing  object  type   DATABASE _ EXPORT/
PLUGTS _ FULL/FULL/PLUGTS _ TABLESPACE
Processing  object  type   DATABASE _ EXPORT/
PLUGTS _ FULL/PLUGTS _ BLK
Processing  object  type   DATABASE _ EXPORT/
EARLY _ OPTIONS/VIEWS _ AS _ TABLES/TABLE _ DATA
Processing  object  type   DATABASE _ EXPORT/
NORMAL _ OPTIONS/TABLE _ DATA
Processing  object  type   DATABASE _ EXPORT/
NORMAL _ OPTIONS/VIEWS _ AS _ TABLES/TABLE _
DATA
Processing  object  type   DATABASE _ EXPORT/
SCHEMA/TABLE/TABLE _ DATA
Total estimation using BLOCKS method: 58.25 MB
Processing  object  type   DATABASE _ EXPORT/
PRE _ SYSTEM _ IMPCALLOUT/MARKER
Processing  object  type   DATABASE _ EXPORT/
PRE _ INSTANCE _ IMPCALLOUT/MARKER
```

```
Processing    object    type    DATABASE _ EXPORT/
   TABLESPACE
Processing object type DATABASE _ EXPORT/PROFILE
Processing    object    type    DATABASE _ EXPORT/
   SYS _ USER/USER
Processing    object    type    DATABASE _ EXPORT/
   SCHEMA/USER
Processing object type DATABASE _ EXPORT/ROLE
Processing object type DATABASE _ EXPORT/GRANT/
   SYSTEM _ GRANT/PROC _ SYSTEM _ GRANT
Processing object type DATABASE _ EXPORT/SCHEMA/
   GRANT/SYSTEM _ GRANT
Processing    object    type    DATABASE _ EXPORT/
   SCHEMA/ROLE _ GRANT
Processing    object    type    DATABASE _ EXPORT/
   SCHEMA/DEFAULT _ ROLE
Processing    object    type    DATABASE _ EXPORT/
   SCHEMA/TABLESPACE _ QUOTA
Processing    object    type    DATABASE _ EXPORT/
   RESOURCE _ COST
Processing    object    type    DATABASE _ EXPORT/
   TRUSTED _ DB _ LINK
.................................................
........ ...........
Master    table    "SYSTEM"."SYS _ EXPORT _ FULL _ 01"
   successfully loaded/unloaded
***********************************************************
************** ********

Dump file set for SYSTEM.SYS _ EXPORT _ FULL _ 01
   is:
/tmp/orcl.dmp
***********************************************************
************** ********

Datafiles required for transportable tablespace
   DATA _ TS:
/u01/app/oracle/oradata/orcl/data01.dbf
Datafiles required for transportable tablespace
   EXAMPLE:
/u01/app/oracle/oradata/orcl/example01.dbf
Job "SYSTEM"."SYS _ EXPORT _ FULL _ 01" completed
   with 3 error(s) at Mon Nov 2 09:27:56 2015
   elapsed 0 00:07:04
```

4. All the necessary database files are moved to Oracle Database 12c's folder.

```
[oracle@upgrade    orcl]$    cp    /u01/app/oracle/
    oradata/orcl/data01.dbf        /u01/app/oracle/
    oradata/IMC12C/
[oracle@upgrade    orcl]$    cp    /u01/app/oracle/
    oradata/orcl/example01.dbf     /u01/app/oracle/
    oradata/IMC12C/
```

5. Specific tablespaces are changed to read–write mode on database, and 11g database is also closed.

```
[oracle@upgrade orcl]$ sqlplus / as sysdba
SQL*Plus: Release 11.2.0.4.0 Production on Mon
    Nov 2 09:38:11 2015
Copyright (c) 1982, 2013, Oracle. All rights
    reserved.
Connected to:
Oracle Database 11g Enterprise Edition Release
    11.2.0.4.0 - 64bit Production
With the Partitioning, OLAP, Data Mining and
    Real Application Testing options
SQL> ALTER TABLESPACE data _ ts READ WRITE;
Tablespace altered.
SQL> ALTER TABLESPACE example READ WRITE;
Tablespace altered.
SQL> shu immediate
Database closed.
Database dismounted.
ORACLE instance shut down.
SQL> exit
Disconnected from Oracle Database 11g Enterprise
    Edition Release 11.2.0.4.0 - 64bit Production
With the Partitioning, OLAP, Data Mining and
    Real Application Testing options
```

6. Finally, data is imported to Oracle 12c database. Before this operation, environment variables should be set on Linux.

```
[oracle@upgrade bin]$ export ORACLE _ SID=imc12c
[oracle@upgrade bin]$ ORAENV _ ASK=NO
[oracle@upgrade bin]$ . oraenv
The Oracle base remains unchanged with value /
    u01/app/oracle
[oracle@upgrade bin]$ ORAENV _ ASK=YES
[oracle@upgrade bin]$ sqlplus / as sysdba
```

```
SQL*Plus:  Release  12.1.0.2.0  Production  on  Tue
   Nov 3 03:43:02 2015
Copyright  (c)  1982,  2014,  Oracle.  All  rights
   reserved.
Connected to:
Oracle Database 12c Enterprise Edition Release
   12.1.0.2.0 - 64bit Production
With the Partitioning, OLAP, Advanced Analytics
   and Real Application Testing options
SQL> CREATE OR REPLACE DIRECTORY TEMP _ DIR AS
   '/tmp/';
[oracle@upgrade  bin]$  impdp  system  full=Y
   directory=TEMP _ DIR  dumpfile=orcl.dmp  log-
   file=impdporcl.log  transport _ datafiles=  '/
   u01/app/oracle/oradata/IMC12C/data01.dbf',  '/
   u01/app/oracle/oradata/IMC12C/example01.dbf'
Import:  Release  12.1.0.2.0 - Production  on  Mon
   Nov 2 11:46:26 2015
Copyright  (c)  1982,  2014,  Oracle  and/or  its
   affiliates. All rights reserved.
Password:
Connected to: Oracle Database 12c Enterprise
   Edition Release 12.1.0.2.0 - 64bit Production
With the Partitioning, OLAP, Advanced Analytics
   and Real Application Testing options
Master  table  "SYSTEM"."SYS _ IMPORT _ FULL _ 01"
   successfully loaded/unloaded
Starting      "SYSTEM"."SYS _ IMPORT _ FULL _ 01":
   system/********  full=Y  directory=TEMP _ DIR
   dumpfile=orcl.dmp       logfile=impdporcl.log
   transport _ datafiles=/u01/app/oracle/oradata/
   IMC12C/data01.dbf,      /u01/app/oracle/oradata/
   IMC12C/example01.dbf
Processing  object  type  DATABASE _ EXPORT/
   PRE _ SYSTEM _ IMPCALLOUT/MARKER
Processing  object  type  DATABASE _ EXPORT/
   PRE _ INSTANCE _ IMPCALLOUT/MARKER
Processing  object  type  DATABASE _ EXPORT/
   PLUGTS _ FULL/PLUGTS _ BLK
Processing  object  type  DATABASE _ EXPORT/
   TABLESPACE
```

```
ORA-31684:  Object  type   TABLESPACE:"UNDOTBS1"
   already exists
Job "SYSTEM"."SYS _ IMPORT _ FULL _ 01" completed
   with 587 error(s) at Mon Nov 2 12:10:40 2015
   elapsed 0 00:24:09
```

7. Consequently, a query is executed for checking all steps of the upgrade process.

```
[oracle@upgrade bin]$ sqlplus test/test
SQL*Plus: Release 12.1.0.2.0 Production on Mon
   Nov 2 12:16:44 2015
Copyright (c) 1982, 2014, Oracle. All rights
   reserved.
Connected to:
Oracle Database 12c Enterprise Edition Release
   12.1.0.2.0 - 64bit Production
With the Partitioning, OLAP, Advanced Analytics
   and Real Application Testing options
SQL> select count(*) from t1;
COUNT(*)
----------
86955
```

MANUAL UPGRADE

Although DBUA is a popular method for upgrading, Manual Upgrade, upgrade with scripts, could be seen as an alternative method. The following steps are necessary in this process to upgrade the database.

1. Execute preupgrd script for checking system prerequisites before upgrade.

```
SQL> @preupgrd
Loading Pre-Upgrade Package...
***********************************************************
*************** *****
Executing Pre-Upgrade Checks in ORCL...
***********************************************************
*************** *****  ****************************
***************************
====>> ERRORS FOUND for ORCL
```

The following are *** ERROR LEVEL CONDITIONS ***
that must be addressed
prior to attempting your upgrade.
Failure to do so will result in a failed upgrade.
You MUST resolve the above errors prior to
upgrade
****
****
****
****

====>> PRE-UPGRADE RESULTS for ORCL
ACTIONS REQUIRED:
1. Review results of the pre-upgrade checks:
/u01/app/oracle/cfgtoollogs/orcl/preupgrade/
preupgrade.log
2. Execute in the SOURCE environment BEFORE
upgrade:
/u01/app/oracle/cfgtoollogs/orcl/preupgrade/
preupgrade _ fixups.sql
3. Execute in the NEW environment AFTER upgrade:
/u01/app/oracle/cfgtoollogs/orcl/preupgrade/
postupgrade _ fixups.sql
****
****
****
************** *****
Pre-Upgrade Checks in ORCL Completed.
****
************** *****
****
************** *****
****
************** *****

2. Execute preupgrade_fixups script.
   SQL>   @/u01/app/oracle/cfgtoollogs/orcl/preup-
   grade/preupgrade _ fixups.sql
   Pre-Upgrade Fixup Script Generated on 2015-11-
   02 07:53:33 Version: 12.1.0.2 Bui ld: 006
   Beginning Pre-Upgrade Fixups...
   Executing in container ORCL
   ****
   ****

```
Check Tag: DEFAULT _ PROCESS _ COUNT
Check Summary: Verify min process count is not
  too low
Fix Summary: Review and increase if needed,
  your PROCESSES value.
*********************************************************
*************
Fixup Returned Information:
WARNING: --> Process Count may be too low
Database has a maximum process count of 150
  which is lower than the
default value of 300 for this release.
You should update your processes value prior to
  the upgrade
to a value of at least 300.
For example:
ALTER SYSTEM SET PROCESSES=300 SCOPE=SPFILE
or update your init.ora file.
*********************************************************
*************
*********************************************************
*************
Check Tag: EM _ PRESENT
Check Summary: Check if Enterprise Manager is
  present
Fix Summary: Execute emremove.sql prior to
  upgrade.
*********************************************************
*************
Fixup Returned Information:
WARNING:  --> Enterprise Manager Database
  Control repository found in the database
In Oracle Database 12c, Database Control is
  removed during
the upgrade. To save time during the Upgrade,
  this action
can be done prior to upgrading using the
  following steps after
copying rdbms/admin/emremove.sql from the new
  Oracle home
- Stop EM Database Control:
$> emctl stop dbconsole
```

```
- Connect to the Database using the SYS account
  AS SYSDBA:
SET ECHO ON;
SET SERVEROUTPUT ON;
@emremove.sql
Without the set echo and serveroutput commands
  you will not
be able to follow the progress of the script.
************* Fixup Summary ***********
4   fixup   routines   generated   INFORMATIONAL
  messages that should be reviewed.
*************** Pre-Upgrade Fixup Script Complete
  ********************
```

3. Apply the following recommended settings on the database.
```
SQL>ALTER SYSTEM SET PROCESSES=300 SCOPE=SPFILE;
SQL>SET ECHO ON;
SQL>SERVEROUTPUT ON;
SQL>@     /u01/app/oracle/product/12.1.0.2/db _ 1/
  rdbms/admin/emremove.sql
SQL>@ ?/olap/admin/catnomad.sql
SQL> EXECUTE dbms _ stats.gather _ dictionary _
  stats;
PL/SQL procedure successfully completed.
SQL> SHUTDOWN IMMEDIATE;
Database closed.
Database dismounted.
ORACLE instance shut down.
```

4. Move spfile and password file from Oracle Database 11g's folder to Oracle Database 12c's folder.
```
[oracle@upgrade dbs]$ cp spfileorcl.ora /u01/
  app/oracle/product/12.1.0.2/db _ 1/dbs/
[oracle@upgrade dbs]$ cp orapworcl /u01/app/
  oracle/product/12.1.0.2/db _ 1/dbs/
```

5. As you are almost done with configuration, set new environment variables.
```
[oracle@upgrade db _ 1]$ export ORACLE _ SID=orcl
[oracle@upgrade db _ 1]$ ORAENV _ ASK=NO
[oracle@upgrade db _ 1]$ . oraenv
The Oracle base remains unchanged with value /
  u01/app/oracle
[oracle@upgrade db _ 1]$ ORAENV _ ASK=YES
```

```
[oracle@upgrade db _ 1]$ sqlplus / as sysdba
SQL*Plus: Release 12.1.0.2.0 Production on Mon
   Nov 2 08:29:05 2015
Connected to an idle instance.
SQL> STARTUP UPGRADE;
ORACLE instance started.
Total System Global Area 613797888 bytes
Fixed Size 2255712 bytes
Variable Size 457180320 bytes
Database Buffers 150994944 bytes
Redo Buffers 3366912 bytes
Database mounted.
Database opened.
SQL>exit;
```

6. In Oracle Database 12c's folder, execute the following script in admin directory.

```
[oracle@upgrade  ~]$  cd  $ORACLE _ HOME/rdbms/
admin
[oracle@upgrade   admin]$   $ORACLE _ HOME/perl/
   bin/perl catctl.pl catupgrd.sql
Argument list for [catctl.pl]
SQL Process Count n = 0
SQL PDB Process Count N = 0
Input Directory d = 0
Phase Logging Table t = 0
Log Dir l = 0
Script s = 0
Serial Run S = 0
Upgrade Mode active M = 0
Start Phase p = 0
End Phase P = 0
Log Id i = 0
Run in c = 0
Do not run in C = 0
Echo OFF e = 1
Grand Total Time: 1518s
LOG FILES: (catupgrd*.log)
Upgrade Summary Report Located in:
/u01/app/oracle/product/12.1.0.2/db _ 1/cfgtool-
   logs/imc12c/upgrade/upg _ s ummary.log
Grand Total Upgrade Time: [0d:0h:25m:18s]
```

7. After the execution of the script, take a look at the upgrade log.

```
[oracle@upgrade db _ 1]$ vi /u01/app/oracle/prod-
uct/12.1.0.2/db _ 1/cfgtoollogs/imc12c/upgrade/
upg _ s ummary.log
JServer JAVA Virtual Machine VALID 12.1.0.2.0
  00:00:00
Oracle Real Application Clusters OPTION OFF
  12.1.0.2.0 00:00:01
Oracle Workspace Manager VALID 12.1.0.2.0 00:00:00
OLAP Analytic Workspace VALID 12.1.0.2.0 00:00:00
Oracle OLAP API VALID 12.1.0.2.0 00:00:00
Oracle Label Security VALID 12.1.0.2.0 00:00:00
Oracle XDK VALID 12.1.0.2.0 00:00:00
Oracle Text VALID 12.1.0.2.0 00:00:00
Oracle XML Database VALID 12.1.0.2.0 00:00:00
Oracle Database Java Packages VALID 12.1.0.2.0
  00:00:00
Oracle Multimedia VALID 12.1.0.2.0 00:00:00
Spatial VALID 12.1.0.2.0 00:00:00
Oracle Application Express VALID 4.2.5.00.08
  00:00:00
Oracle Database Vault VALID 12.1.0.2.0 00:00:00
Final Actions 00:00:41
Post Upgrade 00:07:13
Total Upgrade Time: 00:24:11
PL/SQL procedure successfully completed.
Elapsed: 00:00:00.20
Grand Total Upgrade Time: [0d:0h:25m:18s]
```

8. After examining the log, execute utlu121s script.

```
[oracle@upgrade admin]$ sqlplus / as sysdba
SQL*Plus: Release 12.1.0.2.0 Production on Mon
  Nov 2 12:47:22 2015
Copyright (c) 1982, 2014, Oracle. All rights
  reserved.
Connected to an idle instance.
SQL> startup
ORACLE instance started.
Total System Global Area 616562688 bytes
Fixed Size 2927384 bytes
Variable Size 507512040 bytes
Database Buffers 100663296 bytes
Redo Buffers 5459968 bytes
```

```
Database mounted.
Database opened.
SQL> @utlu121s.sql
PL/SQL procedure successfully completed.
PL/SQL procedure successfully completed.
CATCTL REPORT = /u01/app/oracle/product/12.1.0.2/
  db _ 1/cfgtoollogs/imc12c/upgrade/upg _ s
  ummary.log
PL/SQL procedure successfully completed.
Oracle Database 12.1 Post-Upgrade Status Tool
  11-02-2015 12:48:03
Component Current Version Elapsed Time
Name Status Number HH:MM:SS
Oracle Server UPGRADED 12.1.0.2.0 00:16:11
JServer JAVA Virtual Machine VALID 12.1.0.2.0
  00:00:00
Oracle Real Application Clusters OPTION OFF
  12.1.0.2.0 00:00:01
Oracle Workspace Manager VALID 12.1.0.2.0 00:00:00
OLAP Analytic Workspace VALID 12.1.0.2.0 00:00:00
Oracle OLAP API VALID 12.1.0.2.0 00:00:00
Oracle Label Security VALID 12.1.0.2.0 00:00:00
Oracle XDK VALID 12.1.0.2.0 00:00:00
Oracle Text VALID 12.1.0.2.0 00:00:00
Oracle XML Database VALID 12.1.0.2.0 00:00:00
Oracle Database Java Packages VALID 12.1.0.2.0
  00:00:00
Oracle Multimedia VALID 12.1.0.2.0 00:00:00
Spatial VALID 12.1.0.2.0 00:00:00
Oracle Application Express VALID 4.2.5.00.08
  00:00:00
Oracle Database Vault VALID 12.1.0.2.0 00:00:00
Final Actions 00:00:41
Post Upgrade 00:07:13
Total Upgrade Time: 00:24:11
PL/SQL procedure successfully completed.
SQL>
SQL> --
SQL> -- Update Summary Table with con _ name
  and endtime.
SQL> --
```

```
SQL>    UPDATE    sys.registry$upg _ summary    SET
   reportname = :ReportName,
2 con _ name=SYS _ CONTEXT('USERENV','CON _ NAME'),
3 endtime = SYSDATE
4 WHERE con _ id = -1;
1 row updated.
SQL> commit;
Commit complete.
```

9. Finally, execute post-upgrade script.

```
SQL>      @/u01/app/oracle/cfgtoollogs/orcl/preup-
   grade/postupgrade _ fixups.sql
SQL> REM Post Upgrade Script Generated on:
   2015-11-02 08:09:30
SQL> REM Generated by Version: 12.1.0.2 Build:
   006
SQL> SET ECHO OFF SERVEROUTPUT ON FORMAT
   WRAPPED TAB OFF LINESIZE 750;
PL/SQL procedure successfully completed.
```

10. Check the status with sqlplus.

```
[oracle@upgrade admin]$ sqlplus / as sysdba
SQL*Plus: Release 12.1.0.2.0 Production on Mon
   Nov 2 13:01:56 2015
Copyright (c) 1982, 2014, Oracle. All rights
   reserved.
Connected to:
Oracle Database 12c Enterprise Edition Release
   12.1.0.2.0 - 64bit Production
With the Partitioning, OLAP, Advanced Analytics
   and Real Application Testing options
SQL> SELECT name, open _ mode FROM v$database;
NAME OPEN _ MODE
--------- --------------------
ORCL READ WRITE
```

SUMMARY

This chapter discusses about database upgrade. All of the upgrade methods discussed help you during the upgrade process. In addition, more details can be obtained from the Oracle database web site and its documents page.

Exadata Software Installation

12

Oracle Exadata is entirely an application for managing all related systems that contain network, hardware, operating system, storage software, etc.; thus, for this system is not enough to know only hardware or software. It helps customers by performing better in its tasks. Oracle understands that providing support to customers in installation, patch, and upgrade is valuable in developing customer relations; thus, an insight into customer needs is important for providing support. Exadata needs to be installed with all its structures from scratch just like other systems; it needs some steps to be performed like the installation of operating systems. Exadata installation can be performed basically in eight steps, but all these steps involve Exadata equipment such as network, operating system, and the database that the customer needs; thus, many features can be changed based on customer preference. Installation of Exadata is not easier than that of other systems because it involves all the hardware and software installed in the systems, and moreover, the method of installation of Exadata is slightly unique. Exadata first needs hardware installation, and after purchasing the hardware, it must be installed particularly with an elastic configuration; otherwise, it doesn't need any hardware installation. Hardware installation is critical to better understand Exadata. As this book does not provide the details of hardware installation for Exadata, they can be got from web sites including support.oracle.com. This book focuses on software and database installation through tools in Exadata. Exadata installation is not easy; thus, first of all, the documents in Oracle support web page must be read before beginning the installation, because if a mistake is made during installation, it must be restarted. Therefore, more time should be spent on reading the notes in the support page more carefully (document no. 888828.1). Generally this number is assigned by a technical guy because all the details related to Exadata are stored with this number in the support page.

Note: The following installation procedure is for an old version, so refer to the support note for your Exadata version; otherwise, this installation will be a failure.

First of all, some steps need to be followed during installation; for instance, during installation, if some problems are encountered, a service request needs to be created using the Oracle support page, but the products aren't assigned. So the installation must be completed, and then the below steps must be performed.

1. Enter support.oracle.com on the web browser.

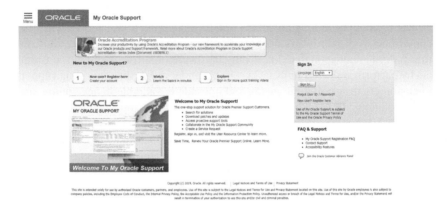

2. Enter your support credentials: your username and password; by the way, if you don't have an account, you need to create one. (It is compulsory.)

3. A new account can be created with the "Create Account" button.

4. Every hardware has a number; check this number on the "My Account" page, and navigate support identifiers.

5. If there is no specific hardware, the hardware must be requested from the owner and then assigned a number.

Now, take a look at how to configure Oracle Exadata.

1. Enter support.oracle.com on the web browser.

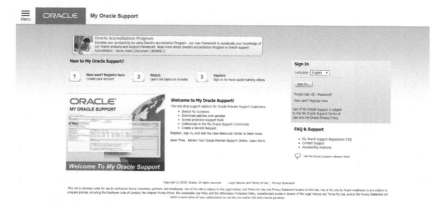

2. Enter your support credentials: your username and password.

3. Write 888828.1 in the right search bar on the support dashboard.

4. This page has all details of the engineered system; so, it is really important for installers. By the way, it shows many new patches related to engineered systems.

5. First find the Oracle exadata deployment assistant (OEDA) file on this page. OEDA is a configuration tool to install Exadata; particularly this tool serves to configure the network, system, and database. If this configuration is on Microsoft Windows, download Windows. We download the desired version of OEDA and the operating system and create a config file and then run it on the Exadata node.

> Power Distribution Unit (PDU) metering unit
>
> Oracle Exadata Deployment Assistant (OEDA)
>
> ?ferences

6. Select the version that will be installed on the system.

Oracle Exadata Deployment Assistant (OEDA)

OEDA deployment tool is used to perform initial Oracle Exadata Database Machine configuration. OEDA deployment tool loads network settings, creates user accounts, installs Oracle Database software, and secures the system based on information in the configuration files created by OEDA configuration tool. Refer to Exadata documentation for details.

Review OEDA README for requirements of versions selected for installation.

OneCommand Release	Versions Supported	Notes
Nov 2018 v181130 - Patch 28972371	18 (18.1.0-18.4.0) 12.2.0.1 (BP170620-RU181016) 12.1.0.2 (BP1-BP181016) 11.2.0.4 (BP1-BP181016)	
Oct 2018 v181016 - Patch 28762834	18 (18.1.0-18.4.0) 12.2.0.1 (BP170620-RU181016) 12.1.0.2 (BP1-BP181016) 11.2.0.4 (BP1-BP181016)	Exadata 19 support
Sep 2018 v180912 - Patch 28665798	18 (18.1.0-18.3.0) 12.2.0.1 (BP170620-RU180717) 12.1.0.2 (BP1-BP180831) 11.2.0.4 (BP1-BP180717)	
Aug 2018 v180821 - Patch 28519707	18 (18.1.0-18.3.0) 12.2.0.1 (BP170620-RU180717) 12.1.0.2 (BP1-BP180717) 11.2.0.4 (BP1-BP180717)	
Jul 2018 v180730 - Patch 28321000	18 (18.1.0-18.3.0)	

7. Click the link for the file to be used.

8. Before the download, click the "Read Me" button, and read all the details related to this patch.

Oracle® Exadata Database Machine and Oracle SuperCluster README for Oracle Exadata Deployment Assistant Update 26619798
11*g* Release 2 (11.2), 12c Release 1 (12.1) , and 12c Release 2 (12.2)

Oracle® Exadata Database Machine and Oracle SuperCluster

README for Oracle Exadata Deployment Assistant Update 26619798

11*g* Release 2 (11.2), 12c Release 1 (12.1) , and 12c Release 2 (12.2)

August 2017

This document describes how to apply the Oracle Exadata Deployment Assistant update. It contains the following topics:

- Section 1, "About Oracle Exadata Deployment Assistant"
- Section 2, "Prerequisites for Oracle Exadata Deployment Assistant Deployment Tool"
- Section 3, "About Oracle Exadata Deployment Assistant Deployment Tool"
- Section 5, "Documentation Accessibility"
- Section 4, "Changes in Previous Releases"

1 About Oracle Exadata Deployment Assistant

Oracle Exadata Deployment Assistant includes a configuration tool, and a deployment tool. It uses the configuration file created by the configuration tool to configure the Oracle Exadata system.

Oracle Exadata Deployment Assistant configuration tool runs on a client. The client must run one of the following operating systems:

- Oracle Linux x86-64
- Oracle Linux SPARC (64-bit)
- Oracle Solaris x86-64 (64-bit)
- Oracle Solaris SPARC (64-bit)
- Microsoft Windows
- Apple OS X (64-bit)

9. Click the "Download" button, and download the file.

10. The message "Downloading now ..." appears on the page.

p26619798_122112....z... ∧
126/172 MB, 39 secs left

11. Click the "config.cmd" file and execute it as shown in the screenshot below. Then click the "Next" button.

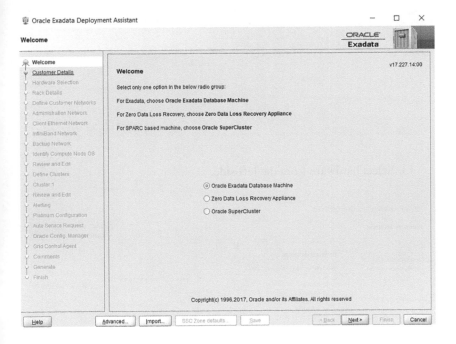

12. Please fill the following necessary information as shown below.
    Network Domain Name: This information is related to the domain name registered in the Domain Name Servers (DNS).
    NTP: This information is needed to access a network time protocol (NTP) service.
    DNS: This information is needed for a specific DNS server address.

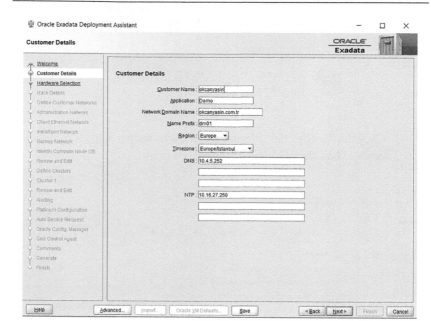

13. Select hardware from the left side.

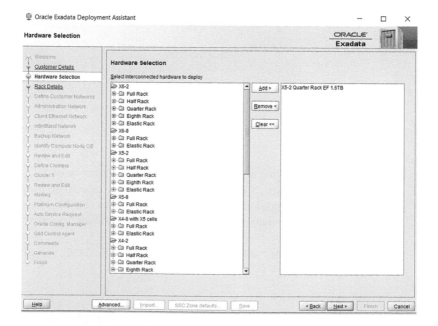

14. Specify hardware prerequisite on the system. As it is known, Exadata has "compute" and "storage cell" nodes.

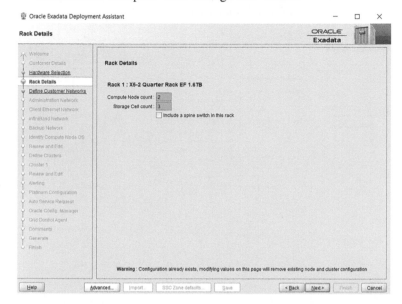

15. Specify subnets for admin and client networks, and these networks are isolated from each other as shown below.

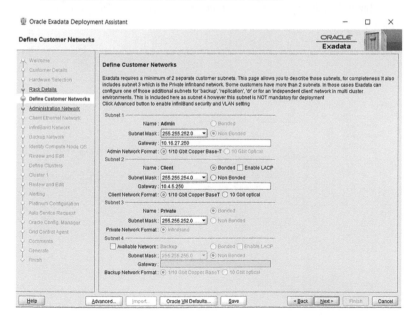

16. Specify the administration network on the OEDA to manage this network in the installation. After that, fill all the details related to "database admin" and "storage admin" nodes.

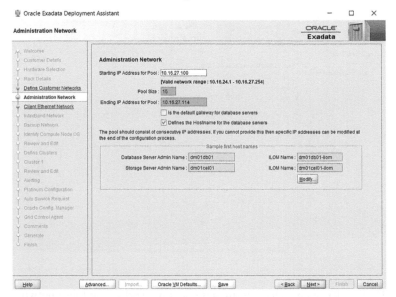

17. Specify the client network on the OEDA to manage this network. After that, fill all the details related to "compute client" and "client scan" nodes.

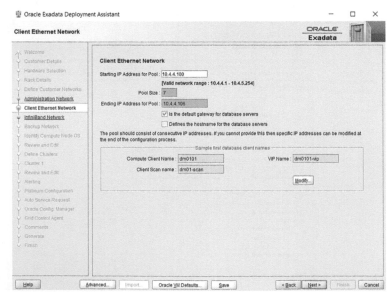

18. Specify the InfiniBand network on the OEDA to manage this network. After that, fill all the details related to "compute priv" and "cell priv" names.

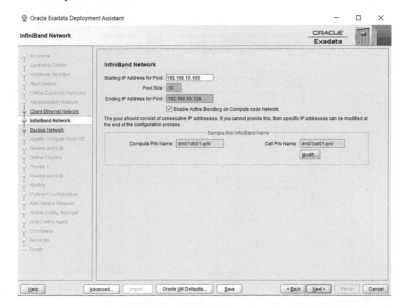

19. If needed, configure a backup network.

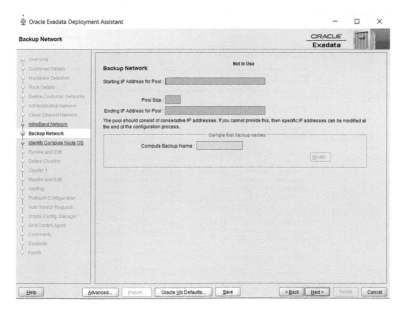

20. Identify your nodes.

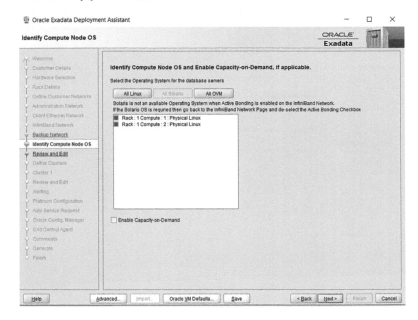

21. Review and edit your network details on this screen.

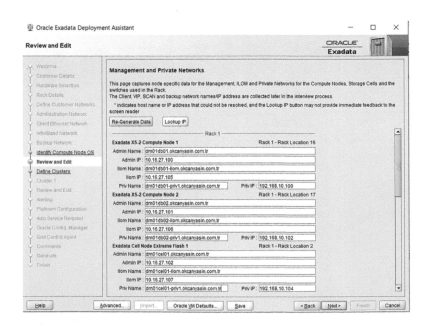

22. Define your cluster.

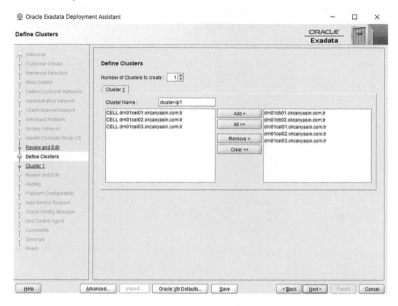

23. Specify Oracle database credential details. Username and group names are needed to create a database. Specify the suitable path to create a database.
24. Review and edit cluster details.

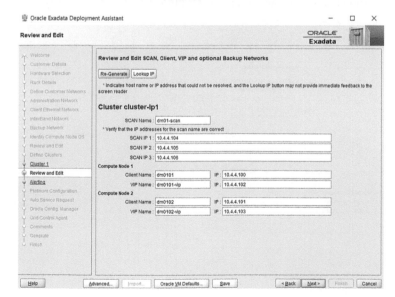

25. If needed, enable the "alerting" service by which the system sends a mail related to the details of the system. Thus, credentials for enabling this service are needed.

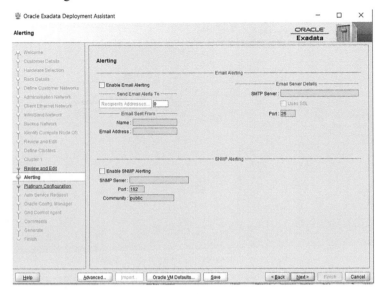

26. If platinum service is needed, configure this service.
27. Oracle Exadata can autocreate a service request for which the nodes shown in the below screen must be filled.

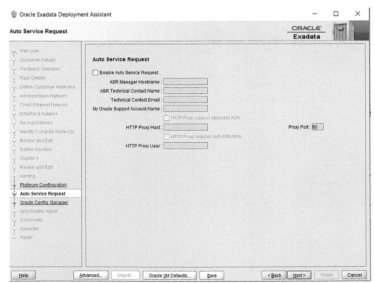

28. Fill config.Manager details if needed.

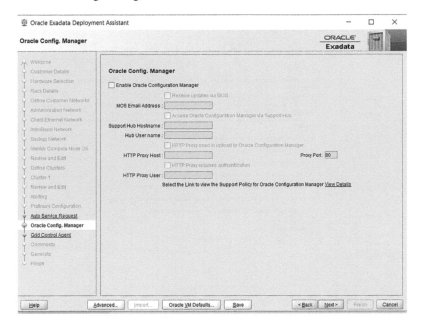

29. Configure grid control agent.

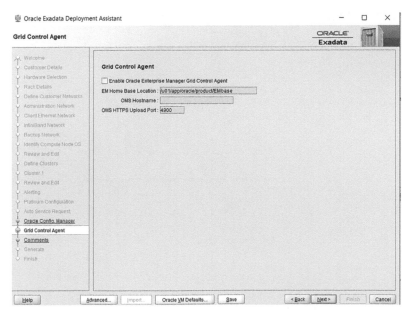

30. Fill the comments page.

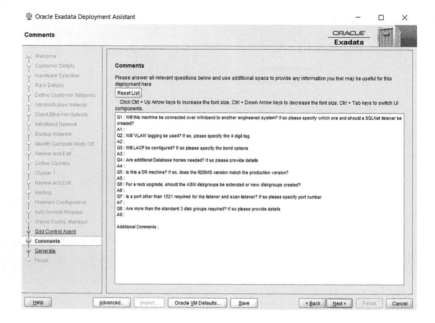

31. Click the "Next" button and generate an XML file.

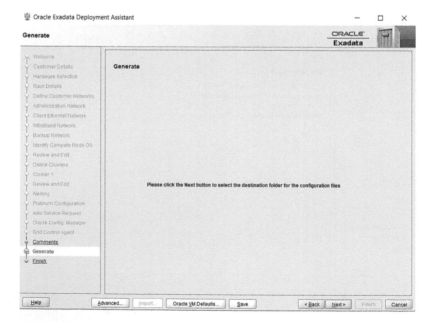

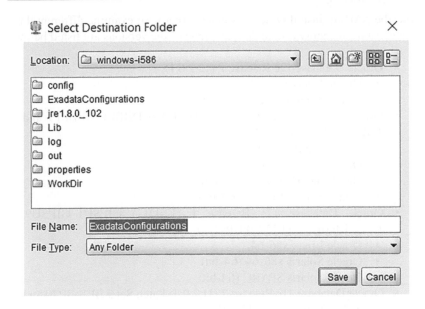

32. Click the "Finish" button and view the XML file available in the specified folder.

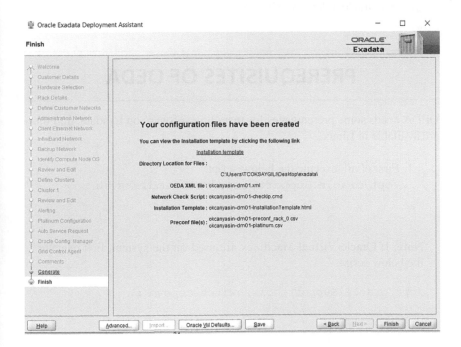

Now, the XML to install Oracle Exadata is ready, so examine different database versions to choose a suitable one. Some of the versions are listed below.

- Oracle Database 12c Release 2 (12.2.0.1)
  - Oracle Linux x86-64
  - Oracle Solaris SPARC (64-bit)
- Oracle Database 12c Release 1 (12.1.0.2) DBBP1–DBBP13 and 12.1.0.2.160119–12.1.0.2.170418
  - Oracle Linux x86-64
  - Oracle Linux SPARC
  - Oracle Solaris x86-64 (64-bit)
  - Oracle Solaris SPARC (64-bit)
- Oracle Database 12c Release 1 (12.1.0.1) GIPSU1–GIPSU9, 12.1.0.1.160119 and 12.1.0.1.160419
  - Oracle Linux x86-64
  - Oracle Solaris x86-64 (64-bit)
  - Oracle Solaris SPARC (64-bit)
- Oracle Database 11g Release 2 (11.2.0.4) Patch Set 3 BP1–BP20 and 11.2.0.4.160119–11.2.0.4.170418
  - Oracle Linux x86-64
  - Oracle Solaris x86-64 (64-bit)
  - Oracle Solaris SPARC (64-bit)

PREREQUISITES OF OEDA

OEDA needs some prerequisites for installing software on Exadata, and these are available in Linux.

1. First of all, execute the below script.
   `/opt/oracle.SupportTools/reclaimdisks.sh`

Note: If Oracle Virtual Machines are used on the system, first execute the below script.

`/opt/oracle.Supporttools/switch_to_ovm.sh`

2. Download OEDA file from support.oracle.com.
3. Execute the below script.

```
unzip -d /u01/onecommand/ p21561234_121111_
Linux.zip
```

4. Specify the WorkDir path after unzip ping the files. The down-loaded files must be placed in the OEDA WorkDir directory.

```
/u01/onecommand/Linux/WorkDir
```

5. Database software can be downloaded from support.oracle.com.
   Oracle Linux x86-64
   V839960-01.zip
   V840012-01.zip
6. OPatch (12.1.0.1.2 or later) release must be downloaded.
7. Patch 19774304 must be downloaded, and this patch is related to engineered systems and in-memory.
8. Execute the install.sh file with rootuser, and the installation will progress step by step. Follow the steps listed below for software installation.
   1. Validate Configuration File
   2. Set Up Required Files
   3. Create Users
   4. Set Up Cell Connectivity
   5. Verify InfiniBand
   6. Calibrate Cells
   7. Create Cell Disks
   8. Create Grid Disks
   9. Configure Cell Alerting
   10. Install Cluster Software
   11. Initialize Cluster Software
   12. Install Database Software
   13. Relink Database with RDS
   14. Create Automatic Storage Management (ASM) Diskgroups
   15. Create Databases
   16. Apply Security Fixes
   17. Set Up Auto Service Request (ASR) Alerting
   18. Create Installation Summary
   19. Resecure Machine

SUMMARY

This chapter explains the installation of Oracle Exadata software but does not cover hardware and operation system installations. Even if software has to be installed in Exadata, it must be done before operation.

References

1. L. LI, B. Niu and X. Tang, "Online Marketing Research Based on Social Choice Theory and Cloud Computing," *2015 IEEE/ACM 8th International Conference on Utility and Cloud Computing (UCC)*, Limassol, 2015, pp. 391–396.
2. O. Y. Saygili, ©2017 Oracle IaaS Quick Reference Guide to Cloud Solutions.
3. B. Grad, "Relational database management systems: The formative years [Guest editor's introduction]," *IEEE Annals of the History of Computing*, vol. 34, no. 4, pp. 7–8, 2012. doi: 10.1109/MAHC.2012.66.
4. https://en.wikipedia.org/wiki/Relational_database_management_system.
5. E. F. Codd, A relational model of data for large shared data banks, *Communications of the ACM*, vol. 13, no. 6, pp. 377–387, 1970. doi: 10.1145/362384.362685.
6. "New Database Software Program Moves Macintosh into The Big Leagues". tribunedigital-chicagotribune. Retrieved 2016-03-17.
7. "A Relational Model of Data for Large Shared Data Banks".
8. SIGFIDET '74 Proceedings of the 1974 ACM SIGFIDET (now SIGMOD) workshop on Data description, access and control.
9. R. Ramakrishnan, D. Donjerkovic, A. Ranganathan, K. S. Beyer, M. Krishnaprasad, "SRQL: Sorted Relational Query Language," *Proceedings of Tenth International Conference on Scientific and Statistical Database Management*, 1998, pp. 84–95.
10. https://en.wikipedia.org/wiki/Relational_database_management_system#/media/File:RDBMS_structure.png.
11. "SQL". Oxforddictionaries.com. Retrieved 2017-01-16; "SQL". Britannica.com. Retrieved 2013-04-02.
12. "SQL Guide". Publib.boulder.ibm.com. Retrieved 2017-01-16.
13. "Structured Query Language (SQL)". Msdn.microsoft.com. Retrieved 2017-01-16.
14. SQL-92, 4.22 SQL-statements, 4.22.1 Classes of SQL-statements "There are at least five ways of classifying SQL-statements", 4.22.2, SQL statements classified by function "The following are the main classes of SQL-statements"; SQL:2003 4.11 SQL-statements, and later revisions.
15. M. Chatham, Structured Query Language by Example - Volume I: Data Query Language. Lulu.com, 2012.
16. E. F. Codd, "A relational model of data for large shared data banks," *Communications of the ACM. Association for Computing Machinery*, vol. 13, no. 6, pp. 377–387, 1970. doi: 10.1145/362384.362685.
17. "Structured Query Language (SQL)". International Business Machines. October 27, 2006. Retrieved 2007-06-10.
18. M. Chapple, "SQL Fundamentals". Databases. About.com. Retrieved 2009-01-28.
19. https://docs.oracle.com/database/121/ADMIN/cdb_mon.htm#ADMIN13720.
20. https://oracle-base.com/articles/12c/rman-table-point-in-time-recovery-12cr1.
21. http://allthingsoracle.com/oracle-database-12c-rman-new-features-part1/.

22. http://www.oracle.com/technetwork/issue-archive/2013/13-sep/o53asktom-1999186.html.
23. https://docs.oracle.com/database/121/SQLRF/statements_7002.htm#SQLRF01402.
24. https://docs.oracle.com/cd/B28359_01/server.111/b28310/tables003.htm.
25. http://www.oracle.com/technetwork/database/in-memory/overview/twp-oracle-database-in-memory-2245633.html.
26. http://oracle4ryou.blogspot.com.tr/2013/05/ora-00845-memorytarget-not-supported-on.html.
27. https://oracle-base.com.
28. https://oracle-base.com/articles/12c/oracle-db-12cr1-installation-on-oracle-linux-7.
29. https://oracle-base.com/articles/12c/multitenant-manage-users-and-privileges-for-cdb-and-pdb-12cr1.
30. https://blogs.oracle.com/imc/entry/sql_translation_framework.
31. https://oracle-base.com/articles/12c/heat-map-ilm-ado-12cr2.
32. https://docs.oracle.com/database/121/BRADV/rcmflash.htm#BRADV89737.
33. http://www.ooug.org/presentations/2014/tom_kyte/12_more_things_about_oracle_12c.pptx.
34. https://docs.oracle.com/javadb/10.8.3.0/devguide/cdevtricks21248.html.
35. http://docs.oracle.com/cd/E80920_01/SAGUG/exadata-storage-server-monitoring.htm#SAGUG20806.
36. https://docs.oracle.com/cd/E24693_01/doc.11203/e15883/exadata.htm.
37. https://www.safaribooksonline.com/library/view/expert-oracle-exadata/9781430262428/9781430262411_Ch04.xhtml.
38. https://docs.oracle.com/cd/E69290_01/doc.44/e71333/concepts.htm#BIGUG107.
39. http://www.orafaq.com/forum/t/186773/.
40. https://blogs.oracle.com/imc/exadata-smartscan-for-ias.
41. https://docs.oracle.com/database/121/TGSQL/tgsql_optop.htm#GUID-461E7071-2229-4F60-82E6-BC4F6FC8D23B.
42. http://docs.oracle.com/cd/E80920_01/SAGUG/exadata-storage-server-monitoring.htm#SAGUG20806.
43. http://docs.oracle.com/cd/E80920_01/SAGUG/exadata-storage-server-iorm.htm#SAGUG20447.
44. https://docs.oracle.com/database/121/ADMIN/tables.htm#ADMIN13948.
45. https://uhesse.com/ 2011/01/21/exadata-part-iii-compression/.
46. http://docs.oracle.com/cd/E80920_01/SAGUG/exadata-storage-server-monitoring.htm#SAGUG20877.
47. http://docs.oracle.com/cd/E80920_01/SAGUG/exadata-storage-server-cellcli.htm#SAGUG20559.
48. http://docs.oracle.com/cd/E80920_01/SAGUG/exadata-storage-server-configuring.htm#SAGUG20368.
49. http://docs.oracle.com/cd/E80920_01/SAGUG/exadata-storage-server-configuring.htm#SAGUG-GUID-C982BD9E-1D55-4F87-A3A0-5897E606652C.
50. http://docs.oracle.com/cd/E80920_01/SAGUG/exadata-storage-server-monitoring.htm#SAGUG20492.
51. http://docs.oracle.com/cd/E80920_01/SAGUG/exadata-storage-server-monitoring.htm#SAGUG20463.

52. http://docs.oracle.com/cd/E80920_01/SAGUG/exadata-storage-server-monitoring.htm#SAGUG20498.
53. http://docs.oracle.com/cd/E80920_01/SAGUG/exadata-storage-server-monitoring.htm#SAGUG20492.
54. http://docs.oracle.com/cd/E80920_01/SAGUG/exadata-storage-server-software-introduction.htm#SAGUG20312.
55. http://docs.oracle.com/cd/E80920_01/SAGUG/exadata-storage-server-cellcli.htm#SAGUG20559.
56. https://blogs.oracle.com/imc/oracle-enterprise-manager-13c-is-available-for-you.
57. https://oracle-base.com/articles/12c/upgrading-to–12c.
58. http://docs.oracle.com/database/121/UPGRD/toc.htm.
59. https://docs.oracle.com/database/121/UPGRD/title.htm.
60. http://www.oracle.com/technetwork/database/upgrade/upgrading-oracle-database-wp-12c-1896123.pdf.
61. http://ioracle-dba.blogspot.com.tr/2015/08/upgrade-oracle-11204-database-to-oracle.html.
62. http://blog.umairmansoob.com/running-exachk-on-exadata-machine/.
63. https://docs.oracle.com/en/engineered-systems/exadata-database-machine/sagug/exadata-storage-server-iorm.html#GUID-CF1C0C2A-7E10-4DB6-8A2B-F217BD1FEC21.
64. https://docs.oracle.com/en/database/oracle/oracle-database/12.2/upgrd/toc.htm.
65. https://docs.oracle.com/database/121/UPGRD/toc.htm.
66. https://www.oracle.com/technetwork/database/upgrade/upgrading-oracle-database-wp-12c-1896123.pdf.
67. https://www.oracle.com/technetwork/database/upgrade/overview/upgrading-oracle-database-wp-122-3403093.pdf.
68. http://www.oracle.com/technetwork/database/exadata/exadataservice-ds-2574134.pdf.
69. https://www.oracle.com/technetwork/database/exadata/exadata-dbmachine-x4-twp-2076451.pdf.
70. https://bdrouvot.wordpress.com/2015/04/30/direct-path-read-and-enq-ko-fast-object-checkpoint/.
71. https://www.oracle.com/technical-resources/articles/enterprise-manager/exadata-commands-part3.html.

Index